ACKNOWLEDGEMENTS

KU-077-291

The 1975 manual (no. 2) was an enlarged and completely revised version of the Friends of the Earth Whale Campaign Manual, published in June 1972. It has, for this (1978) printing, been updated, includes the most recent figures for catches and quotas, and incorporates some of the changes made necessary by the FAO Scientific Consultation on Marine Mammals, held in Bergen, Norway in September 1976.

The original was written by Nic Holliman, Angela King, Stephanie Lenz and Graham Searle. It was revised in 1975 by Tom Burke, John A. Burton, Joanna Gordon Clark, Angela King, Rob Page, David Pedley, Angela Potter, Brian Price, Richard Sandbrook and Peter Wilkinson. The 1978 updating was done by Peggy Jones, Jon Barzdo and Joanna Gordon Clark. It was re-typed by Pat Mear.

We should like to thank the staff at the library of the Natural History Museum, Dr. Peter Purves for help with locating original data, the many national and international FoE groups (especially Warwick Fraser (South Africa), Peter Poynton (Australia), George Dewez (France), Rowan Taylor (New Zealand), Bill Orme (Ireland), Lennart Daleus (Sweden), Tom Garrett and Anne Wickham (USA) and other whale conservation groups for information and help. Lastly we wish to thank Dr. Sidney Holt for his invaluable advice on the scientific interpretation of the data.

We are most grateful to the following for their financial support:

The World Wildlife Fund for $3,360.00
Christine Stevens (on behalf of the Animal Welfare Institute) for $500
John Aspinall for £200.

Cover designed by Reg Boorer, from a painting by Larry Foster.

SUMMARY OF THE SECTIONS

Chapter 1   aims to answer any queries the reader may have about the necessity to save whales. Many of the points touched on in this section are dealt with in greater detail in subsequent sections.

Chapter 2   describes the evolution, characteristics, life cycles and place of whales in the ecosystem.

Chapter 3   deals with the varying efforts of governments, administrators and politicians to save whales. It outlines the failures of the IWC*, particularly in respect of providing adequate protection for currently exploited species. This section stresses that political action is the only way in which whales will be saved.

Chapter 4   gives an historical account of whaling legislation and of the International Whaling Commission.

Chapter 5   itemises the activities of the whaling nations (apart from Japan).

Chapter 6   is concerned with the economics of whaling; it reviews the present situation and outlines the options open to the industry.

Chapter 7   deals with the scientific aspects of whaling and underlines the uncertainties in the methods used by scientists to determine the population sizes of many species of whales.

Chapter 8   describes the methods which are used to kill whales.

Chapter 9   the action guide: this aims to give the reader some guidelines as to how he or she can personally and actively participate in determining the future of whales.

The Appendices   include detailed information on the currently exploited and protected species and small species of whales, alternatives to whaling in the Antarctic, whale products and alternatives to sperm oil.

International Whaling Commission

"This Commission will be known to history as a small body of men who failed to act responsibly in the terms of a very large commitment to the world and who protected the interests of a few whalers and not the future of the world's whales."

The Mexican delegate to the 1974
meeting of the International
Whaling Commission

CONTENTS

INTRODUCTION

In 1946 an international governmental organisation was established to conserve from overexploitation the whales of the world. The International Whaling Convention included in its schedule the protection of the already overexploited Grey, Right and Antarctic Humpback Whales from member countries. Twelve years later its members caught 2,394 Humpback and over 1,100 Blue Whales in one year. In 1960 the International Whaling Commission (IWC) refused protection to the Blue Whale and over 1,700 were caught. Within the following 5 years the IWC discovered that both the Blue and the Humpback Whales were not only, like the Grey and Right, commercially extinct, but teetering on the brink of disappearance from this planet forever.

By 1976, thirty years after its foundation, it was abundantly clear that the Commission had still not fulfilled its obligations. In that year it was necessary to ban the killing of all Fin Whales in the Southern Hemisphere and North Pacific, Sei Whales in the Nova Scotia region of the North Atlantic and in two regions of the Antarctic, Bryde's Whales in the Southern Hemisphere, Minke Whales in the East and Central South Pacific, and Sperm Whales in areas of the Southern Hemisphere.

Of the 150 member countries of the United Nations, only 17 belong to the IWC, so that approximately 90% of the UN nations have no part in the decisions of the Commission. If whales belong to anyone they belong to everyone and it is essential to the future survival of this planet that all peoples be consulted in the conservation of whales and all other things living in international waters.

Without the work of committed scientists, individuals and organisations from all parts of the world, the plight of the whale would probably be worse — if that is possible — than it now is. However, while the facts are not known, people cannot influence events, and it is the purpose of this Manual to help to spread the information both to a wider public and to those in a position to make influential decisions.

Friends of the Earth are calling for a moratorium on all commercial whaling for a minimum of ten years to prevent the extinction of currently overexploited whale species, to allow the depleted stocks of whales to recover, and to give scientists time in which to make proper and accurate assessment of whale populations and their ecology. We call too for far greater funding of non-destructive research.

The UK Government should:

1) Introduce an import ban on all Sperm Whale products and especially sperm oil. The Sperm Whale constitutes 40% by weight of the total world catch, and the

Antarctic stocks have decreased in size from 54 to 45 feet and in weight from 47 to 27 tons from 1933 to 1972. Unless measures are taken soon to conserve Sperm Whales, their position becomes as precarious as that of the Blue and Humpback Whales (see Appendix 4).

2) *Persuade the United Nations members to declare whales the common heritage of mankind.*

3) *Act through international organisations to ensure that any exploitation of the living resources of the Antarctic (e.g. krill, whales, fish, seals), if and when it occurs is managed by one truly international body, with conservation as a first priority, and with a brief to ensure that the benefits are equitably distributed.*
Adopt a complementary stance at the Law of the Sea Conference.
Seek control over the taking of small whales.

4) *Initiate import bans on all whale products in the UK and in the EEC and press for overt support for the 10 year moratorium within the EEC (see Chapter 9).*

5) *Put pressure on the USSR and Japan to phase out whaling while negotiations for the 10 year moratorium continue (see Chapter 9).*

6) *Influence non-IWC whaling countries with which we have close political, economic or traditional ties (e.g. Portugal) to join the IWC, and pressure IWC member countries to refuse to trade in whale products with non-member nations.*

We want individuals to:

1) *Put pressure on their elected representatives to ensure that these objectives are achieved.*

2) *Exert their own personal pressure by boycotting goods containing whale derivatives.*

3) *Put pressure on the British leather industry to replace sperm whale oil with substitutes, by refusing to buy leather goods until sperm oil is banned from the UK.*

NOTE

This 1978 edition is an updated version of the 1975 Whale Campaign Manual 2. It includes the up to date figures for catches and quotas, and incorporates some of the changes made necessary by the FAO Scientific Consultation on Marine Mammals, held in Bergen, Norway in September 1976.

The major part of this updating was done by Peggy Jones, with contributions from Jon Barzdo and Joanna Gordon Clark.

All references quoted are more fully detailed in the Bibliography, Appendix 14, page 134.

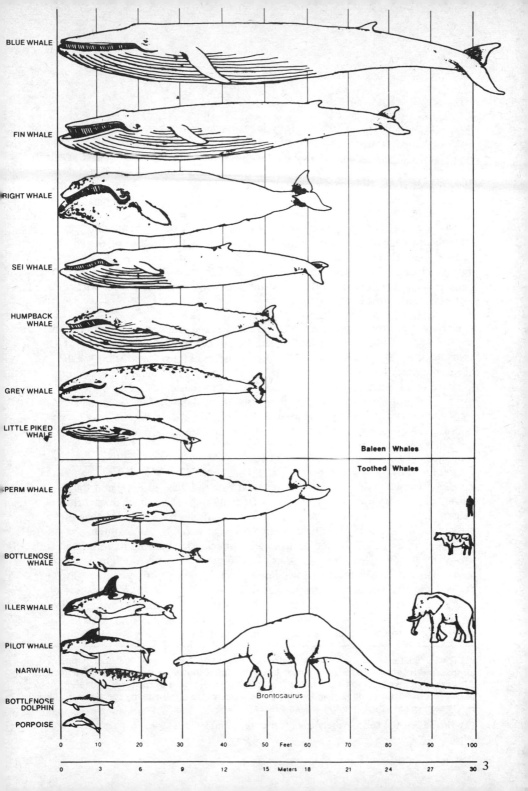

BLUE WHALE

FIN WHALE

RIGHT WHALE

SEI WHALE

HUMPBACK
WHALE

GREY WHALE

LITTLE PIKED
WHALE

Baleen | Whales

Toothed | Whales

SPERM WHALE

BOTTLENOSE
WHALE

ILLER WHALE

PILOT WHALE

NARWHAL

BOTTLENOSE
DOLPHIN

PORPOISE

Brontosaurus

| 0 | 10 | 20 | 30 | 40 | 50 Feet 60 | 70 | 80 | 90 | 100 |

| 0 | 3 | 6 | 9 | 12 | 15 Meters 18 | 21 | 24 | 27 | 30 |

3

CHAPTER 1

WHY SAVE WHALES?

When people talk of "the whale" they are usually thinking of one of the species of "great" whales such as the Blue or the Humpback. What is often misunderstood is that there are about 80 species of whales, dolphins and porpoises.[1] These species vary tremendously in distribution, eating habits, behaviour and size (from 4 or 5 feet to over 100). The main characteristics which they have in common are that they are all completely aquatic mammals and breathe air through their lungs.

When we talk of saving whales, we are talking not only of saving the Blue or the Humpback Whale, but of conserving and maintaining the balance of the whole marine ecosystem in which the whales play a significant part. It is, of course, important to conserve all the living resources of the seas. But the whales, perhaps, are a special case in that they are one of the few resources in international waters that are already exploited.

Whales evolved from mammals that lived on the land and have gradually become superbly adapted to aquatic life.[2] The ability to live in a wide range of habitats, is one of the main reasons for the biological success of a species. Some whales live in the polar seas and the equatorial oceans; others have adapted to live in fresh water. It is not known why whales moved to the sea, but "the sea does offer an environment that is richer in nourishment than the land" (Man's Place). Apart from the Dugong and Manatee which breed in coastal waters, whales are the only mammals which spend all their life in the sea and do not even come onto the land to breed.

Whales are the largest creatures ever to have inhabited the world. They can be divided into two main groups — the toothed whales (Odontocetes) and the baleen whales (Mysticetes) which eat by filtering their food through vast fibrous baleen plates. Recent studies of some species of whales have revealed that they are gentle, and probably intelligent creatures with strong social instincts. Scientists have also discovered that the Humpback Whale is able to communicate over vast distances by means of 'songs'. Undoubtedly there is much more we can learn about, and probably from, whales, but only so long as they remain alive. Perhaps the growing concern for whales reflects man's admiration of success — perhaps it is simply that whales have become a symbol of man's growing awareness of his environment, but that there is a growing concern on the part of large numbers of people all over the world cannot be denied.

(1) It was "well into the 20th century before cetology left the realm of wild speculation and approached reality. Even today our knowledge of whales shows great gaps on very elementary subjects . . . There are species of whales that have only been seen in the form of badly preserved stranded individuals, and we have a very poor idea of how many specimens of these rarely seen species there may be". (Man's Place)

(2) This adaptation to an entirely aquatic life is so complete that for a long time man thought whales were fish.

4

Of all the species which comprise the order Cetacea, we suspect that the toothed Sperm Whales are the most successful. They are more varied and more numerous than the baleen whales. They do not have any predators (except man), and are equipped with excellent brains.[1] These are thought to be the largest and best developed of all living creatures.[2]

Other toothed whales such as the Killer Whale and some species of dolphins have shown a marked propensity for learning complex tricks in captivity. Sperm Whales are gregarious, care for their young and breeding females, and unlike baleen whales, are polygamous. Also, unlike the baleen whales, or any other toothed whales as far as is known, Sperm Whales are the only species which have conquered both the depth and the breadth of the oceans.

We do not feel that man has either the right or the need to destroy any species or habitat at will. Our ignorance of the systems with which we are interfering is profound. The long term outcome of our actions is largely unknown. Man, as the most successful of all species, has special responsibilities. The decisions he takes now affect not only his own and his children's future, they also affect that of many other species. In this situation the over-exploitation of whales becomes as immoral as it is insensitive.

A more selfish reason for wanting to conserve whales is that if we conserve whale stocks now, we may well ensure an important supply of protein for future generations — when meat protein is certain to be scarce. Therefore every year we delay the implementation of the 10 year moratorium, the more we reduce the possibility of being able to feed starving people in the future.[3] It is not only in our interest to ensure that all whale species survive man's onslaught, but it is our duty to preserve and foster the considerable food resources that whales could provide if properly managed, for the time when the words 'food' and 'survival' become synonymous. If harvested properly it has been calculated that whales could provide the world with two million tons of protein annually. This resource is all the more significant if one remembers that the total world fish harvest cannot be relied upon to solve the world's food problems. In fact, some

---

(1) Our limited knowledge of whales restricts our capacity to understand the ways in which their intelligence expresses itself. We simply cannot assume that they are less intelligent than we are.

(2) "The human brain weighs 1 400 grams, and the dolphin's 1 700. The dolphin's cerebral cortex is larger than ours. It has twice the number of convolutions, and 10 to 40% more nerve cells ... On balance, then, it looks as though the dolphin has more cortex left over for the higher mental processes than we do." (Man's Place)

(3) "If the stock of fin whales had been managed properly it could have supplied man in perpetuity with an annual sustainable yield of 20 000 whales. Twenty thousand fin whales a year lost to hungry mankind means a loss of over 1 million tons of cetacean raw material, roughly twice as much as the blue whale. Without going into further statistical detail the loss to mankind of the large whales in the Southern Hemisphere is a tragedy involving the destruction of a perpetual source of food for over 12 million human beings. Man may have a higher I.Q. than his cetacean cousins but a superiority in common sense is much in doubt." (The Blue Whale, George Small, p205)

important fish stocks have also been seriously over-exploited, and the overall catch is no longer increasing in the way that it has been for the past 20 - 30 years.

Concern for the future of the whales is not new. The whaling industry itself realised in the 1930's that international agreement was necessary in order to stop the over-exploitation of the resource. What is new, however, is the growing symbolic importance of whales. As Dr. Robert White, the Commissioner for the USA to the IWC recently said, "whales and their management today are a symbol of a much broader concern for the preservation of his environment". This concern, manifested in the form of political pressure, must continue if the whales are to survive.

CHAPTER 2

THE BIOLOGY OF WHALES

a)   EVOLUTION

About 100 million years ago the ancestors of whales left the land and adapted to a completely aquatic environment. Whales are thought to have descended from an early group of terrestrial carnivorous mammals called creodonts, from which many mammals such as dogs, cats and seals are also descended.

There are very few fossil remains of primitive whales, so very little is known about how (or where) the change from the land to the sea took place. The earliest fossil of a whale found is estimated to be 55 million years old and had teeth resembling those of the early creodonts. It is thought that the reason why earlier fossil remains of whales have not been found is because when the ancestors of whales began to inhabit the sea they were in areas where fossils were rarely formed, and that when they became completely aquatic, the fossils were again rarely formed in the open sea.

Remains of an early whale were found in 1845. It was snake-shaped and 55 feet long. The flipper had already evolved from the fore-limb but the elbow was still apparent. Many other fossils have been found since then, and they have varied from the serpent-shaped type described above to those resembling modern porpoises.

b)   CHARACTERISTICS OF WHALES

Since whales began to live in the sea, all parts of their bodies have become enormously modified to make them suited to their aquatic environment. For example, the hind limbs have disappeared, the fore limbs have become flippers, they have evolved horizontally flattened tail flukes and the nostrils have moved to the top of the head. Some of the functions of hair have been replaced by blubber and, in the case of baleen whales, the functions of teeth have been replaced by horny plates. Navigation by sight has been replaced by navigation using sound (sonar). These modifications are explained in more detail below:

1)   **Size:** the most obvious characteristic of the great whales, only possible by the adoption of a marine environment, is their very size (a Blue Whale weighing 140 tons is equivalent to 30 average sized elephants). Without the buoyancy bestowed by water, no animal could attain these gigantic proportions.

2)   **Heat Retention:** apart from a few whiskers around the snout, whales have lost the characteristic mammalian hair which in land mammals plays an

important role in insulation. Instead, hair is replaced by fat (blubber) which varies in thickness from about 1 inch in porpoises to over 1 foot in Sperm and Right Whales. Blubber also provides the whales with a reserve of food and, perhaps, water. In the case of the Sperm Whale which can dive to great depths, it is thought that the blubber may act as a resistance against pressure. The fat tissue is sufficiently compressible to permit cetaceans to streamline their body shape during swimming, in order to cut down turbulence.[1]

3) **Modification of the Limbs**: the forelimbs have been modified into a pair of rigid flippers which are situated behind the head. They can only be bent at the shoulder joint and act like paddles. The arm bones are very short and the five fingers are enclosed in fibrous tissue. Some of the faster swimming species of whales also have a dorsal fin in the middle of the back which acts as a stabiliser. These fins contain no bone structure but are simply folds of skin. The hind limbs are entirely absent except for one or two detached bones on each side of the body which are not visible externally. In the place of hind limbs, whales have developed a strong tail and two enormous horizontal tail flukes projecting on either side which beat up and down, unlike the tails of fish which move from side to side, and which provide great propulsion. It is the upward thrust which pushes the whale forward by displacing water upwards and backwards. These tail flukes, like dorsal fins, have no bony support inside.

4) **Feeding**: the baleen, or whalebone, whales, lack teeth but have enormous sieve-like plates (the baleen or whalebone) which hang from the upper jaw and which reach lengths of over 10 feet in the Greenland Right Whale. These plates are used to filter the zooplankton (mainly krill) and fish, on which they feed from the water. To feed, the whale opens its mouth and swims into a shoal of krill. When the mouth is full, the whale closes its mouth and by raising its tongue, pushes the water out through the baleen plates. The krill is trapped by the baleen and swallowed. The toothed whales, which include the dolphins and porpoises, Killer Whale and Sperm Whale, feed mainly on fish, squid and cuttlefish. The Killer Whale is a true predator, hunting small whale species, seals and penguins.

5) **Breathing**: breathing takes place through a blowhole which opens upwards and which is situated on the top of the head and sunk below the general surface of the skull. The rapidly filled and rapidly emptied lungs allow long periods of submergence (Sperm Whales have been known to stay

---

(1) "The bodies of cetaceans are not perfectly streamlined but they are living animals, not rigid bodies . . . The way the skin of a cetacean is fastened to the underlying blubber allows minute changes in the shape of the surface to be made so that eddies are stopped as soon as they begin to appear, turbulence is prevented and laminar flow is preserved. The inner surface of the skin is covered with dermal pupillae, or ridges, that are believed to function in these adjustments by which drag is reduced so much as to be practically absent." Harrison – Matthews, "The Whale".

submerged for a maximum period of over one hour). Unlike man, whales manage to resist the effects of compression when they dive deeply, particularly the 'bends' which are caused by nitrogen in the blood. There are various reasons why whales can do this: firstly, very little air is kept in the lungs when diving (thus reducing the amount of nitrogen that can be absorbed), secondly, what little nitrogen that is released on surfacing is absorbed by oil in the blowhole side passages and is expelled in the form of foam when the whale blows.

6) **Head**: the head region (which is asymmetrical in the toothed whales) shows the greater modification. The front and back parts of the skull have been 'telescoped' in the course of evolution, whereas the mid-skull has been greatly enlarged to accommodate the brain and to provide support for the enormous jaws. The eyes are small and are flush with the side of the head. They are adapted for seeing under water, but are not nearly such important sense organs as the ears, which are very small at the surface, but the inner end of the ear passage is large.

Only in the last few decades has man begun to understand the methods by which whales navigate. They use sonar by emitting clicks and squeaks and, like bats, dolphins (and probably whales) are able to navigate accurately from the echoes produced by their sounds. It has also been found that cetaceans are capable of some form of communication and this field is being researched at present. The function of the sonorous songs of the Humpback Whale is not yet known.

7) **Reproductive Organs**: the male reproductive organs are normally contained within the body, thereby avoiding heat loss and retaining the streamlined contouring. Similarly, the female mammary glands empty into grooves rather than taking the form of the enlarged breasts of most other mammals.

c) THE LIFE CYCLE OF WHALES

We are only just beginning to understand the fabric of cetacean society. Recently it was discovered that whales are able to communicate with each other over enormous distances.[1] This discovery has revealed another dimension to their already complex and mysterious existence. Many whales are gregarious, take great care of their young, help the females in labour and protect the mothers with calves by keeping them safe in the middle of the herd.

Most great whales migrate to warmer waters in winter to breed and return to colder polar waters during the summer months to feed. These migrations are often of distances over 5 000 miles each way. The whales' gestation periods differ for each species: baleen whales' pregnancies are estimated to last from

---

(1) "It has been suggested that the excellent conduction of sound through water allows whales to hear each other over distances of hundreds of miles." Robert Burton, "The Life & Death of Whales", p.65.

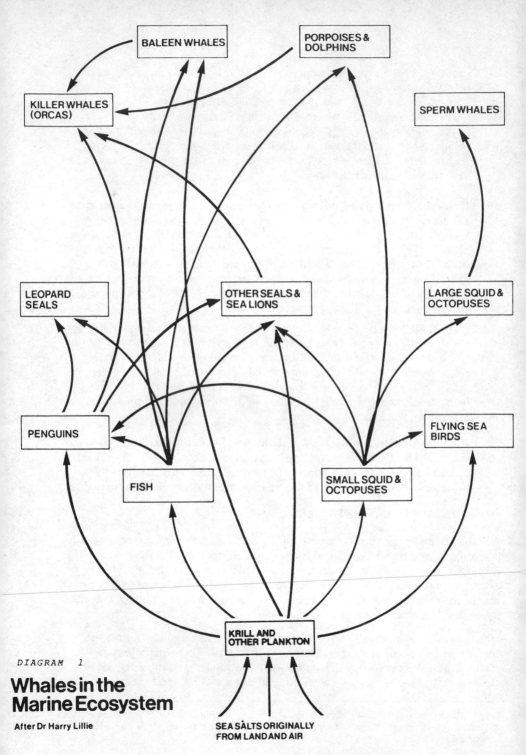

BALEEN WHALES

PORPOISES & DOLPHINS

KILLER WHALES (ORCAS)

SPERM WHALES

LEOPARD SEALS

OTHER SEALS & SEA LIONS

LARGE SQUID & OCTOPUSES

PENGUINS

FLYING SEA BIRDS

FISH

SMALL SQUID & OCTOPUSES

KRILL AND OTHER PLANKTON

DIAGRAM 1

Whales in the Marine Ecosystem

After Dr Harry Lillie

SEA SALTS ORIGINALLY FROM LAND AND AIR

about 10 to 12 months. Baleen whales are monogamous and breed once every two years. The young calves are large and well-developed at birth. For example, a Blue Whale calf is 25 feet long and weighs over two tons. It doubles its weight within a week (its mother's milk is very rich) and adds 200 pounds to its weight every day. The weaning period usually lasts six to seven months.

The breeding cycle of Sperm Whales is slightly different from that of the baleen whales. The Sperm Whale is polygamous. Mating takes place in the spring and the pregnancy lasts 14 to 16 months — as it also does in the Killer Whale. The young are weaned after about 2 years. The length of the breeding cycle means that the Sperm Whales can only breed once every 4 to 5 years.

The average life expectancy of unexploited great whales is estimated at 60 to 90 years.

d)   WHALES IN THE ECOSYSTEM

Whales play an important role in the life of the ocean. Baleen whales are at the top of a short food chain beginning with the diatoms, which float in the upper layers of the sea and which convert the sun's energy and salts dissolved in the sea into food for the next link in the chain — the krill. Krill (a collective term for zooplankton* on which baleen whales feed) varies in its constituents in different parts of the world. In the southern oceans, krill consists almost exclusively of one species of prawn-like crustaceon called Euphausia superba, whereas in northern waters a type of mollusc called the sea butterfly provides the whales' staple diet. There seems to be a delicate relationship between the distribution of whales and plankton, balanced not only by the eating habits of the whale, but perhaps also by its excretion, which has been seen to give rise to great population explosions amongst planktonic organisms.

The toothed whales form the top of a longer food chain which can be simplified to a diatom (phytoplankton) — zooplankton — fish/squid — Sperm Whale progression (see Diagram1). Killer Whales have been seen to take sickly whales as well as penguins and seals. It can be argued that in this way they tend to play the same part that wolves, seals and otters play in other ecosystems in keeping the prey species healthy.

As the total living mass (biomass) of whales decreases, more nutrients and energy pass to the remaining organisms. The result can be an increase in their populations (in our terms often a pest problem) and a generally less stable environment. This has been the consequence on a number of other occasions when natural predators or organisms high in the food chain have been eliminated from formerly more stable ecosystems. A similar example with some practical implications is provided by birds of prey, rodents and a field of grain. The farmer's grain is the food of some rodent "pests", and the rodents are the food of some birds of prey.

---

*From Greek words meaning 'animals which drift'.

The number of birds are controlled by the numbers of rodents available as food and the numbers of rodents are controlled by the availability of grain. A similar example with readily apparent results is provided by birds of prey, rabbits and downland. The downlands were kept open and grassy (once sheep had been removed) by rabbits. The number of buzzards and some other predators was very closely linked to the rabbit populations. When myxomatosis wiped out rabbit populations, the numbers of buzzards declined as they were unable to find alternative food supplies. Scrub encroached on the downland and when the rabbit populations began to recover, the available grassland had decreased or disappeared and so it was impossible for them to return to the former situation.

The destruction, or even continued reduction, of whale populations could therefore trigger off greater problems than have yet to become apparent to man. Whaling has already caused large disturbances in the Antarctic ecosystem. Sei Whales and Minke Whales are now believed to be maturing younger, so also are crabeater and fur seals. This implies that for all these animals the rate of reproduction has gone up, and in the case of the fur seals there has been a marked increase in populations. Penguins are also thought to be increasing. These changes are all believed to stem from a greater abundance of krill than existed when the ecosystem was in a balanced state before Antarctic whaling began. Now that Blue and Fin Whales — the largest consumers — have been so heavily reduced, it is supposed that more krill is available to other animals. There is no indication at present that Blue Whales are recovering, despite 12 years of protection. Many scientists think it extremely unlikely that severely depleted whale species, such as the Blue, Humpback and Right will ever recover to their former levels.

Every animal and plant species on earth performs unique and specific functions intricately woven into the proper functioning of the whole ecosystems on which our survival depends. Every extinction of a species, every simplification of natural diversity reduces the stability and resilience of ecosystems and makes them (and hence us) more vulnerable to the ordinary accidents of nature. Perhaps the world can get along without whales; perhaps not; but it is not the sort of thing one wants to find out empirically. The nearest analogy we do know about concerns the ecological stability of grasslands, which collapses if one removes the large herbivores. Whales are the marine analogy — they are the great grazers of plankton. If, having killed all whales, we find the world will not work well without them after all, it will be too late. We cannot create whales.

CHAPTER 3

THE POLITICS OF WHALING

Only the political importance given to the whales can save them. However, we cannot expect the politicians to assign the whales this status without considerable and continuous pressure from the general public. The work of many pressure groups and individuals over the past few years has done much to focus public attention on the plight of the whales, but constant vigilance is essential in order to prevent the whale becoming a pawn in the political manoeuverings of an increasingly resource-short world.

The United Nations and its appropriate agencies — the United Nations Environment Programme (UNEP) and the Food and Agriculture Organisation (FAO) — the EEC, the IWC and national governments can all take action to deal with the problem. In the final analysis, however, it is the extent to which national governments are prepared to accept and maintain binding international agreements that will determine the fate of the whales. This in turn is dependent on the work of pressure groups and individual action within nations.

International Efforts to save Whales:

a) THE UNITED NATIONS AND THEIR AGENCIES

One of the first persons to moot the concept that marine resources outside national waters are the "common heritage of mankind" was Arvid Pardo, Ambassador from Malta to the United Nations. At that time, Pardo was talking about the sea bed, but since then (1967) this idea has been applied to other resources of the high seas — such as tuna and whales.

b) UNITED NATIONS CONFERENCE ON THE HUMAN ENVIRONMENT 1972

In the light of this concept, the following resolution was proposed and passed by an overwhelming majority including the UK and USA (53 votes to 0 with 12 abstentions) at the Stockholm Conference in 1972:

"It is recommended that Governments agree to strengthen the IWC, to increase international research efforts, and as a matter of urgency, to call for an international agreement under the auspices of the IWC to involve all governments concerned in a ten year moratorium on commercial whaling."

The Japanese put down an amendment making the 10 year moratorium relate to "depleted" stocks alone, but it was defeated. (Brazil and South Africa voted in favour of the amendment; the USSR did not attend the Conference.) The moratorium was later endorsed by the Governing Council of the United Nations Environment Programme meeting in Geneva in 1973 and again in 1974. UNEP also gave £200,000 to a FAO working party whose remit is to produce an independent assessment of whale populations and other marine mammals (see page 62). In 1974 UNEP sent an observer to the IWC for the first time. (It is interesting to note that the IWC has always voted not to come under the auspices of the UN.)

c)   LAW OF THE SEA CONFERENCE

It has not been decided how the United Nations Law of the Sea Conference (still being negotiated) will affect the whaling industry, but laws based on the "common heritage" principle would mean a change of concept from treating the living resources of the high seas as a private property resource to a common property resource. As an international resource, the management of the whale stocks must be undertaken by a body with a broader membership base than that which is currently represented in the IWC. In effect, whales are a resource that 'belong' to all nations — even though all nations may not possess the technology to kill whales, or have a coastline or wish to use whale products. Even without ecological considerations, the current exploitation of whales by two major nations for their short-term benefit is unacceptable, and we must work to ensure that the Law of the Sea Conference will set up the machinery and enforcement mechanisms to make equitable distribution a reality.

d)   THE INTERNATIONAL WHALING COMMISSION

The IWC is one example of an organisation which is already concerned with the management of an international marine resource. We feel there are many useful lessons to be learnt from the mistakes which have been made by the IWC. Some of these are summarised below.

Failures of the IWC (See page 18 for History of the IWC)

Many of the weaknesses of the IWC date from the structural faults in the original proposals.

1.   One of the most fundamental weaknesses of the Commission has been its failure to act quickly enough. For example, in 1949 the Commission set a total catch limit for Antarctic pelagic whaling at an excessively high quota of 16 000 units, but the IWC was unable to reduce this limit substantially until 1965. As a result, the Blue and Fin Whale stocks were soon decimated in the Antarctic. The usual time-lag of about ten years before which action is taken is clearly no longer acceptable.

Another example of the IWC's slowness to react was its failure to get the Blue Whale unit (BWu) abolished and separate quotas for each species established, until 1972 — even though the Special Committee of Three Scientists recommended to the IWC in 1963 that the BWu should be abolished.

The BWu was based on the approximate yield of oil which the baleen whales in the Antarctic yielded, i.e. 1 Blue Whale = 2 Fin Whales or 2½ Humpbacks or 6 Seis. As the BWu did not afford protection to individual species, the stocks of rarer whales were given no particular protection. The preferred species were the largest, so the Blue, Fin, Humpback and Sei were decimated in turn by the Norwegians, British, Dutch, South Africans, Japanese and Russians. The whalers then turned their attention to whale species which were not included in the BWu — the Sperm and the Minke. If this leisurely pace is continued by the decision makers the whales will not survive.

2. The IWC has no powers of enforcement. Moreover, under the 90 day rule of the Commission any nation can opt out of any of the decisions made at the meetings within the subsequent 90 days. This loophole has frequently been used (see History of IWC, page 18).

3. Although the IWC deals with an international resource it is not an international body. Not all of its 17 member nations[1] whale and some countries which do whale are not members of the Commission. Non-member whaling nations take a significant proportion of the world Sperm Whale catch.

4. Even though the value of the primary products of whales for the few years up to 1975 has been about US $200 million, the annual budget was only about 0.04% of this. Until 1976 there was never any full time staff. The IWC's scientific committee is composed of people who may only spend a fraction of their time studying whales and even then the scope of their work is restrained by the demands of the industry.

5. The failure to get data submitted regularly is one reason why scientific advice may fall short of what is needed in a particular situation. The improved scientific knowledge of the resource was one of the reasons why the Antarctic whaling nations accepted a substantial reduction of the catch.

6. The currently exploited species are being over-exploited in exactly the same way that the Blue, Fin and Humpback were before them. Because of the lack of knowledge and the demands of the industry, the IWC is still allowing the current catch quotas to be much too high.

Governments are complacent about whales because they have failed so far to assess their potential importance in relation to the problems of over population and food shortages. If governments took these latter two points seriously, then the future of the whale stocks would be of supreme political importance. At the moment, their future is essentially determined by two countries and whaling has been, and continues to be, a prime example of short-term thinking. In this context, the success or failure of the IWC also takes on a new significance.

e) THE BRITISH GOVERNMENT'S VIEW OF THE IWC

Although it is praiseworthy that the British Government tries to make international conventions work, there are often problems. In the case of the

---

(1) Argentina, Australia, Brazil, Canada, Denmark, France, Iceland, Japan, Mexico, Netherlands, New Zealand, Norway, Panama, South Africa, UK, USA, USSR.

International Convention for the Regulation of Whaling (see p. 19), the trouble has been that the cost of trying to make it work has been at the expense of the safety of many whale species. Our Government, in common with others, appears to prefer having the machinery established with which to negotiate even when the spirit of the Convention is being violated.

It is questionable whether the International Whaling Commission has done more harm than good. Over the past decade, the IWC has unfortunately given international legitimacy to the pre-determined demands of the Russian and Japanese whaling industries, and other member countries an excuse for inaction. On the positive side, the IWC has given the world a vehicle through which public opinion can be channelled and pressure exerted. It has given interested people a forum in which to discuss the problems of whales and whaling and has also made some information available to the general public which would otherwise have only been available to the industry itself.

The British, and other governments, have taken the view that if too much pressure is put on the Commission that Russia and Japan would walk out and that uncontrolled whaling would be inevitable. However, despite the fact that there is a Commission to regulate whaling, there is still no mechanism whereby countries can be prevented from doing what they want. They do — and will only agree to continue to belong to the Commission if loopholes are available to them.
In 1973, Japan and the USSR opted out of important decisions made at the IWC meeting (see page 32) — quite legally under the rules of the Commission — so the situation would not be that much different if the USSR and Japan did leave. However, we feel that the likelihood of this happening is slim, even though it might seem to be economically attractive in the short term for them to do so. According to the editorial in the Japan Times (July 2, 1974), "Japan in particular would not want to risk economic reprisals by failing to adhere to the IWC's decisions". It follows, therefore, that Japan would be unwilling to leave the Commission even though other countries have walked out at the slightest provocation when they have not liked the decisions made or — in the case of the Netherlands in 1959 — the drift of the conversation.

f) EFFORTS BY NATIONAL GOVERNMENTS

The moratorium is the stated policy of many nations of the world, but with few notable exceptions, it seems that only lip service has been paid to the ideal and that little positive action has been taken. The USA has always been the prime mover behind the moratorium, and was the first country to ban the import of whale products in 1971. Since the Stockholm Conference, the British Government has introduced a ban on most baleen whale imports in March 1973 (see page 81); Canada closed down her whaling stations, Australia banned whale imports from non-IWC countries, New Zealand banned the import of primary whale products, and France is considering such a move.

16

But what have the Governments who voted for the 10 year moratorium at Stockholm done to get the moratorium implemented? The answer is that we do not know for sure, but we suspect very little, and nothing overtly. The USA is the only country to have taken really positive action by adopting economic measures which are designed to put pressure on the Japanese. Under the Fisherman's Protective Act, the Pelly Amendment gave the US Government the powers to cut off fisheries imports from any nations which "act in a manner to diminish the effectiveness of an international conservation organisation".

An amendment to the Pelly Amendment was introduced into the US Congress in 1975, (Magnusson/Dingell, H.R. 15039) to widen the scope of the Pelly Amendment giving the US Government the powers to ban the import of any product from any such country. Thus, if Japan and/or the USSR had opted out of any of the decisions made at the 1974 IWC meeting, the President would have been in a position to ban the import of any Russian or Japanese goods. Both the USSR and Japan are particularly vulnerable to such threats.

But action through exports would also be effective since Japan only produces 53% of her total food requirements and is largely dependent on soya beans from the USA. In 1971, she imported about 3 000 000 tons of soya beans, home production having steadily fallen from 336 000 metric tons in 1962 to only 122 000 metric tons in 1971. When the USA cut off supplies of soya beans in 1972, Japan became even more determined to increase her domestic supply of food-stuffs. The problem with this approach is that the Japanese say, with some justification, that they are therefore forced to increase their domestic supplies of protein by whaling.

A better approach which might be tried by the USA would be to use these tactics in reverse: they might, for instance, guarantee future supplies of soya beans in return for a phase out of whaling, rather than to cut off her exports of soya beans to Japan. Similar tactics could be used for the Soviet Union, but with wheat as the bargaining tool. It is interesting to note that the Russians have said privately many times that they intend to phase out whaling if the Japanese do. This is probably because they are now harvesting krill (see page *139*) directly — without going through the intermediary of the whale. As it is a criminal offence to kill a dolphin in the USSR, because they were over-fished in the Black Sea and are now considered as the marine cousins of man, perhaps the Russians will also afford the same protection to their larger and more over-fished cousins in other waters.

N.B. No action has yet been taken by the EEC on whaling. The reply to a question asked by Lord O'Hagan to the European Parliament on whales was as follows:

"As a result of the experience gained from the implementation of the (Washington) Convention, once it has been ratified, the Commission might, if necessary, propose that measures be taken at community level."
(24th November 1973)

CHAPTER 4

HISTORY OF WHALING LEGISLATION AND
THE INTERNATIONAL WHALING COMMISSION

Even though weaknesses in the IWC's attempts to control whaling exist, its institution represented the most significant step taken in the history of whaling controls.

Pre-IWC Whaling Controls:

1324   approx. For England, Wales and Ireland, De Prerogativa Regis (concerning the Royal Prerogative). Whales, together with sturgeons, were declared "Royal Fish". As such, any caught in the three mile territorial limit or washed up on the beach have to be presented to the monarch. This law was to provide the basis of the extensive collection of cetaceans at the British Museum (Natural History), as the Crown passed all strandings to the Museum.

1821   Russians exclude foreign whalers from Bering Sea.

1904   Norway bans coastal whaling.

1921   British Government introduces licences and restrictions on whaling in Antarctic dependencies.

1929   Norwegians pass whaling act.

1931   Convention for the Regulation of Whaling signed in Geneva.

1935   Limits to catch size introduced under the auspices of the League of Nations.

1936   USA bans killing of Grey Whales.

1937   International Agreement for the Regulation of Whaling signed in London.

1938   Humpback Whales protected in Antarctic.

1939   Atlantic whaling nations except Iceland and Denmark agree not to hunt Blue Whale.

In the past few years, a great deal of criticism has been made of the work of the International Whaling Commission. Nevertheless, when individual Governments of countries which deal either in whaling or in whale products come under fire, they invariably point out that it is really the job of the IWC to conserve whale stocks and that individual Governments support the work of the IWC to the hilt. Yet the IWC itself cannot make decisions. It is currently composed of 17 member nations[1] which together decide the fate of the whales. Thus the IWC is only as good (or bad) as its members.

This section of the Manual traces the history and development of the IWC and is based not on the remarks of critics, but rather on the Reports of the IWC itself, of which there have been 27 and which are publicly available. If we are to place our trust in the IWC, it is important for us to understand its role in the conservation of whales.

1946

In December 1946, the representatives of 14 Governments came together in Washington to sign the International Convention for the Regulation of Whaling. The signatory Governments were Argentina, Australia, Brazil, Canada, Chile, Denmark, France, the Netherlands, New Zealand, Norway, Peru, USSR, UK and the USA.

The purpose of the Convention was to safeguard for future generations "the great natural resources represented by the whale stocks". The document recognised that such conservation was in the common interest and that hitherto "the history of whaling has seen overfishing in one area after another to such a degree that it is essential to protect all species of whales from further overfishing".

The Convention established the International Whaling Commission which was to consist of delegates of all those countries adhering to the Convention; and the original member Governments were asked to decide in the two years preceding the first IWC meeting whether the IWC should become part of a UN Agency.

The Schedule to the Convention laid down the restrictions to be placed on whaling: Right and Grey Whales were given protection, certain areas were out of bounds for the taking of certain whales, the Antarctic pelagic whaling season (with the use of factory ships) was to last from 22nd December — 7th April, minimum lengths for different whales were established, the catching of lactating females and females with calves was prohibited, and the quota system for the taking of Antarctic whales was set at 16 000 Blue Whale units per season. No Humpbacks were to be taken in Antarctic waters, since stocks were so low that the Schedule expressly forbade it. The weaknesses of the original Convention have already been pointed out (see page 14).

1949

The first meeting of the IWC was held in May in London, and one of the first decisions taken was to relax the total ban on the killing of Humpback Whales in the Antarctic and to permit the taking of 1 250 Humpbacks in 1949-50 and in the following season. An amendment to the Schedule was passed, limiting the

---

(1) Membership has continuously fluctuated.

use elsewhere of factory ships normally operating within territorial waters. France became the first nation to use the opting-out procedure contained in Article V, by refusing to be bound by the amendment.

The Convention demanded that amendments of the Schedule should be based on scientific findings and accordingly a Scientific Sub-Committee of the IWC was established. The 1949-50 Antarctic catch by pelagic fleets was 16 059 Blue Whale units — 59 BWu in excess of the quota. 2 117 Humpback Whales were taken — 867 in excess of the permitted limit.

1950

The membership of the IWC, which held its second meeting in Oslo in July, was fifteen. It was decided that the IWC should remain independent of the United Nations.

The IWC ratified the taking of 1,250 Humpback Whales by the Antarctic fleets, and put the Antarctic quota of baleen whales at 16,000 BWu's. The catches for 1950-51 were 16,416 BWu including 1,630 Humpbacks.

1951

The IWC met in July in Cape Town and had seventeen members, Japan and Brazil having joined. The IWC accepted a recommendation of the Scientific Committee to put no minimum length requirement on the taking of Minke Whales, and put a limit of 1 250 on the 1951-52 Antarctic Humpback catch. An attempt was made to place some restriction on the periods in which Sperm Whales could be treated at land stations (an 8 month-1 year season) and the IWC amended the Schedule. But Australia objected and was therefore (under Article V) not bound by the amendment.

In the 1951-52 Antarctic season, 16,007 BWu including 1,545 Humpbacks were taken.

1952

The fourth IWC meeting (held in London in June) set up a Scientific Sub-Committee to examine the state of Humpback stocks in the Antarctic and the 16 000 BWu quota (which was retained for the 1952-53 Antarctic season).

The Antarctic pelagic season and the period in which Humpbacks could be taken were shortened, and the catches were 14,866 BWu including 949 Humpbacks.

1953

The Scientific Sub-Committee reported to the IWC (June in London) that the stocks of Humpbacks were such that a three day Antarctic season should be established for them, that no Blue Whales should be taken before 15th January in any Antarctic season, and that the quota should be reduced from 16 000 BWu to 14 000 BWu. They also argued that the Antarctic Sanctuary for all baleen whales (south of 40°S. from 70°W. to 160°W) should be extended to cover 0° to 70°W. in the case of Blue Whales only, in order to offer relief to this species, stocks of which were low.

In the 1953-54 Antarctic season, 15,456 BWu were taken.

1954

The Report of the Scientific Sub-Committee was presented to the IWC meeting in Tokyo in July. Its recommendations were quite unequivocal: there should be a complete ban on the taking of all Blue and Humpback Whales in the northern hemisphere, there should be no Blue Whales taken in the South Atlantic outside the Antarctic, and there should be a ban on the taking of Blue and Humpback Whales south of 50°S. between 0° and 70°W. Further, the quota for the Antarctic should be reduced from 15 500 BWu to 15 000 BWu.

The IWC received this report via the Scientific Committee and in response banned for five years the hunting of Humpbacks in the North Atlantic and south of 40°S. between 0° – 70°W. It also banned the hunting of Blue Whales for five years in the North Atlantic, but Iceland and Denmark lodged objections and hence the ban did not apply to them. Similarly, Japan, USA, USSR and Canada objected to an IWC ban imposed on hunting Blues in the North Pacific and, consequently, this too had no meaning. The IWC decided not to prohibit the hunting of Humpbacks in the North Pacific, not to restrict the hunting of Blues in the Antarctic, not to ban the hunting of Blues in the South Atlantic, and not to reduce the 15 500 BWu quota for 1954-55.

The IWC did decide to set up another Scientific Sub-Committee to consider some of the points raised at the 6th meeting of the IWC, and concluded that "it may soon become necessary to restrict more severely the Antarctic catch of Blue Whales".

15,323 BWu's including 493 Humpbacks were taken in the 1954-55 Antarctic pelagic season.

1955

The Scientific Committee reporting to the 7th IWC meeting in Moscow (July) concluded that no more than 1 250 Blue Whales should be taken from the Antarctic in the next season. The previous season's catch figures indicated that there was a further decline in the Southern stocks of Blue Whales and that their "powers of recovery might already be found to be largely lost, even if it (the Blue Whale) received total protection".

Southern stocks of Fin Whales, too, were being over-exploited and depleted and the previous season's catch of 26 000 Fins by Antarctic fleets was well in excess of what should have been taken: 19 000 Fins was the absolute maximum (according to the Scientific Committee).

But the Committee did not reiterate its demand for separate species limits for the Antarctic, and only asked that the total quota be reduced to 14 500 BWu. This was what was asked for the year before and since then, that figure had been exceeded by 800 BWu's. 14 500 BWu's (as events proved) was well in excess of what the Southern whale stocks could stand. Furthermore, in an attempt to share the excessive pressure on stocks amongst all the areas of the Antarctic, the Scientific Committee proposed that the only sanctuary hitherto afforded to whales in the Antarctic (south of 40°S. and between 70° and 160°W. – established by the original Schedule to the Convention) should be opened to whaling fleets in 1955-56.

At their July meeting, the IWC decided to rescind the ban on the taking of Blues in the North Pacific: since the North Pacific whaling nations had objected to the ban, it had no meaning anyway. It was decided to open the Antarctic whale sanctuary for three years from November 1955, after which "the area will be automatically closed". Norway raised informally the question of international observers on factory ships and at land stations, but the IWC decided that this was "ultra vires" and did not discuss the matter.

The IWC decided against separate quotas for different species of whales and that the Antarctic quota should be set at 15 000 BWu for 1955-56 and 14 500 for 1956-57. But objections were received to the lowering to 14 500 BWu from the Netherlands, UK, Panama, South Africa, Norway, Japan, USA, Canada and (later) USSR. This latter restriction was therefore not binding on any IWC member with whaling fleets in the Antarctic.

The Convention noted that Chile, Peru and Ecuador had entered into separate arrangements for whaling and since none of these nations was a member of the IWC (although Chile was an original signatory to the Convention), they were not covered by the IWC restrictions on whaling.

At the IWC meeting, Japan accused a Panamanian vessel (the 'Olympic Challenger') of hunting Humpbacks out of season, and produced aerial photographs in an attempt to justify its allegations. Norway too, accused Panama. The Panamanian Commissioner stated that "the Inspectors serving on board the 'Olympic Challenger' were officers of the Panama Government and were competent and trustworthy". The charges were denied, and no record of any infractions (normally sent by all factory ships) were received from the 'Olympic Challenger' in that year.

In the 1955-56 Antarctic season, 14,874 BWu's including 1,432 Humpbacks were reported as taken.

1956

The Scientific Sub-Committee set up at the previous meeting of the IWC reported to the IWC meeting held in July, in London, that there was "no way in which an Inspector can obtain evidence of the killing of nursing females except by the condition of the mammary glands". The condition of the mammary glands could only be established after the female whale had been killed, and hence the ban on the taking of suckling females and females accompanied by a calf, presented great difficulties of enforcement. It was admitted by the 8th IWC that "lactating females can be taken inadvertently", and hence that adherence to the original ban on the taking of suckling females was problematical.

The main concern of the Sub-Committee was that such was the scarcity of Blue Whales in the Antarctic that the Fin Whale stocks were being put under increasing pressure and were suffering depletion. They recommended that the quota for the 1956-57 season should be 14 500 BWu.

At the IWC meeting, it was decided (again) that it was impractical to set specific limits on the taking of different sorts of whales in the Antarctic; (again) that the International Observer Scheme could not be discussed; and that the 1956-57 quota should be put at 14 500 BWu, to revert to 15 000 BWu in 1957-58. The 1956-57 Antarctic catch was 14 745 BWu — 245 BWu in excess of the theoretical maximum permitted.

1957

The Scientific Sub-Committee reporting to the 9th IWC meeting said that "while it could not necessarily be proved that the (Antarctic) Humpback stock had increased", it was their opinion that the season in which Humpbacks could be taken should be lengthened. The reason for this recommendation was not that stocks of Humpbacks would be able to stand increased exploitation, but rather that the more excessive the pressure on Humpbacks, the greater the relief which would be given to over-taxed Fin Whale stocks. The Committee concluded that the Fin Whale stock "is in the process of being seriously depleted" and that "the present number of whales taken annually in the Antarctic is dangerously high". The 15 000 BWu quota should be reduced.

At the IWC meeting, it was decided to reduce the Antarctic quota to 14 500 BWu in 1957-58. However, the catch was 14 850 BWu — 350 BWu in excess of the limit.

1958

At the 10th IWC meeting in The Hague in June, it was decided that the former Antarctic whale sanctuary, which was to have been again closed to whaling fleets "automatically" in November 1958 (and which was not "automatically" closed to whaling for the 1958-59 season) might be closed until 1959-60. This matter was to be reconsidered at the 11th IWC meeting.

The Scientific Sub-Committee proposed an extension of the theoretical five year ban on hunting Blue Whales in the North Atlantic, which was due to expire in 1960. But since Iceland and Denmark were not bound by the existing ban, the IWC decided to defer a decision on this matter. Similarly, due to problems of protocol, the International Observer Scheme could not be decided upon. The ban on the taking of Humpbacks in the North Atlantic was extended by another five years, but the area in which they were to be given protection was reduced.

On the question of the Antarctic quota, there was again some dissension. According to the Schedule, the quota was to revert from 14 500 BWu to 15 000 BWu for 1958-59. The Scientific Sub-Committee had recommended a quota of 14 500 BWu and this was accepted by the IWC and the Schedule was amended accordingly. However, objections were received from the Netherlands, Norway, Japan, USSR and UK, so that none of the nations with whaling fleets in the Antarctic were found by the quota. Instead, they agreed to keep to a 15 000 BWu quota, which in 1958-59, they exceeded by 301 BWu. The catch of 15 301 BWu was 801 BWu in excess of the IWC limit, and included 1 101 Blue Whales and 2 394 Humpbacks (2 000 more than were taken the previous year).

By this time, the IWC was becoming a little sensitive to criticism and consequently decided that verbatim reports of Commission meetings were "not for general public distribution and quotations should not be permitted in public press or in trade journals".

1959

Prior to the 11th IWC meeting in London (June), conferences of the Antarctic pelagic whaling nations were held to allocate national quotas for the 1959-60 season. Agreement could not be reached and Norway and the Netherlands withdrew from the Whaling Convention and the IWC, this taking effect from June 30th 1959. Hence, these two nations were no longer bound by any of the IWC's restrictions on whaling. Japan also submitted notice of withdrawal from the Convention, although after the IWC meeting this was not put into effect.

The official commencement of the 1959-60 Antarctic season was brought forward by ten days and the season lengthened, although the Scientific Committee objected that this would result in the killing of more pregnant females. It was decided that the former sanctuary area, which was to have been closed in 1958, should remain open until November 1962 and progress on neutral observers was blocked by the USSR, which argued that since the Netherlands and Norway would be outside the Convention, the scheme could not work. A working party was set up to examine the question of the humane killing of whales and the case for the use of electric, rather than explosive, harpoons.

In spite of the Scientific Committee's evidence that Antarctic stocks were in serious decline (particularly those of Blue and Fin Whales), the IWC quota for

1959-60 was put at 15 000 BWu. However, since this was not binding on all Governments, each Government set its own limit. These were sufficiently high to make the total quota 17 600 BWu.

The 1959-60 season was a long one and was characterised by a very low rate of catch. There was a marked reduction in the average length of Blue Whales taken and over 50% of these were immature whales. The catch of 15 512 BWu was in excess of any catch in the last seven seasons, and came at a time when every member of the IWC had before him incontrovertible evidence of the depletion of existing stocks.

But worse was to follow. The attitude of the IWC is summed up in the following extract from the Chairman's Report of the 11th meeting: "Conscious of the importance of maintaining the Convention, the Commission showed a willingness to consider making some increase in the Antarctic catch if thereby the loss of those member countries which had given notice of withdrawal could be averted".

It will be remembered that the prime purpose of the Convention, which it was so important to maintain, was to "protect all species of whales from further over-fishing".

1960

An Ad Hoc Scientific Sub-Committee set up the previous year, reported to the IWC meeting held in London. Because stocks of Blue and Fin Whales were showing ever more rapid decline, the Committee argued that the season for taking Blue Whales should be shortened or certain Antarctic regions should be closed to the hunting of Blues. But even such measures as these would be insufficient: "any real recovery of the stock cannot be expected without a period of total protection in the Antarctic". The Committee had a 2-3 year period in mind. Since pressure on Fin Whales was increasing, and more immature whales were being caught, there needed to be a "substantial reduction in total catch".

Humpbacks too were in trouble. The Committee wanted "a complete rest for 10-12 years" in the taking of Humpbacks from the South Atlantic and a total ban on hunting them in one Antarctic region (between 70° and 130°E). Further, the Committee noted "a progressive reduction in the sizes of Sperm Whales from before the war up to the present time", and they regarded this "as a warning that the Antarctic stock was being appreciably affected by whaling".

The IWC refused to give Antarctic Blue Whales protection, and instead passed an amendment whereby Blues could not be hunted before February 14th in any year. Japan, UK and USSR objected to this restriction, Norway rejoined the IWC in time to add her objections to the amendment, and the Netherlands (being outside the IWC) had no need to object. The amendment was, therefore, binding on none of the nations with pelagic fleets in the Antarctic. Humpback whaling was outlawed in one Antarctic region, but no protection was given to Fin, Sei or Sperm Whales. Instead, the IWC decided to set up another Committee, this one to be known as the Committee of Three Scientists to investigate whale stocks and to make recommendations to future meetings of the IWC.

The most important decision taken at the 12th IWC meeting was to invoke absolutely **no quota** for the 1960-61 and 1961-62 Antarctic seasons.

Despite objections from Japan and USSR, the BWu catch quota was suspended. This IWC policy of laissez faire (designed to woo Norway and the Netherlands) resulted in an Antarctic pelagic catch of 16 433 BWu. This was the highest catch figure in the history of the IWC and was never to be repeated, even though all restrictions were removed in the following season too. No less than 71.1% of the 1 740 Blues (including Pygmy Blues) which were killed, were immature whales.

1961

The Report of the Ad Hoc Scientific Sub-Committee presented to the IWC in London, was essentially a reiteration of previous recommendations which had not been acted upon. Blue Whale stocks in the Antarctic "are being given no opportunity of regeneration"; a period of total protection was needed. The stocks of other whales were more critical even than before.

The IWC responded by further lengthening the Antarctic season and by imposing no catch limit whatsoever. The ban on Humpback whaling in part of the Antarctic became meaningless, since all the IWC's Antarctic whaling nations lodged objections to it. Similarly, the shortening of the Blue Whale season was not binding on any of these members or on the Netherlands, which remained outside the IWC. Implementation of the International Observer Scheme was postponed until after the 1961-62 season. The final report of the working party on humane killing was received and the working party disbanded. It was concluded that "explosive harpooning killed as quickly as the electrical method".

The absence of a quota for the Antarctic 1961-62 season enabled the catch to total 15 253 BWu.

1962

The Ad Hoc Scientific Committee reported to the 14th IWC meeting in London, that as long ago as 1955, the Scientific Sub-Committee recommended an Antarctic quota of 11 000 BWu. Whilst this claim is not substantiated by any interpretation of the Sub-Committee report in question (the recommended quota for 1955-56 having been 14 500 BWu), it is true that recommendations of scientific committees had been consistently ignored. The Ad Hoc Committee now recommended a quota of 5 000-7 000 BWu and stated that "If such a quota is not introduced at once, and if the catch is held at the level of recent years, a further and probably accelerating decline in stocks will occur". On the subject of Antarctic Blue Whales, the conclusion was that the "main Blue Whale stocks have now declined to a level of a tenth or less of that immediately after the war, and that the decline shows no sign of slowing down". Nevertheless, the scientists recommended that the former Antarctic sanctuary remain open.

The now 18 members of the IWC (Norway and the Netherlands having rejoined and Argentina joined) decided against all the available evidence to impose a quota of 15 000 BWu for the 1962-63 season. Attempts to reduce the Blue Whale season and to restrict the hunting of Humpbacks were blocked by the objections of the Antarctic whaling nations.

1963

The 15th IWC in London had the benefit of three reports prepared by the Ad Hoc Scientific Committee, the Committee of Three Scientists and the Scientific Sub-Committee. But the recommendations were familiar. The Blue Whale unit should be abolished and replaced by separate species quotas in order to protect the Fin and Sei Whales which were by now bearing the brunt of the attack. There should be a complete ban on the taking of Blue Whales in the Antarctic and Humpbacks in the whole of the southern hemisphere. If such a ban were immediately to be imposed, it would still take stocks of these whales 50 years to build up to a level whereby they could provide the maximum sustainable yield. If a ban on the taking of all Fin Whales were imposed, it would take 5 years for those stocks to show the desired improvement.

The IWC banned the taking of Humpbacks from the southern hemisphere, and banned the taking of Blue Whales in most of the Antarctic. Another attempt was also made to shorten the Antarctic Blue Whale season in the area still open to the hunting of Blues, but this move met with the usual non-compliance of the five whaling nations involved. Predictably, the International Observer Scheme was postponed and the Blue Whale unit was retained.

The quota for 1963-64 was 10 000 BWu, but it became apparent during the season that this target could not be met. The catch included 8 286 Sei Whales (50% more than in the previous season) and totalled 8 429 BWu.

1964

After the 1963-64 Antarctic season, the Scientific Committee pressed for the abolition of the Blue Whale unit and for the establishment of specific limits at the "lowest possible level below 4 000 Fin Whales and 5 000 Sei Whales". All hunting of Blue and Pygmy Blue Whales in the Antarctic should be prohibited. The report of the former Committee of Three Scientists, which by now had become the Committee of Four Scientists, put Antarctic Fin Whale stocks at 32 400 whales, compared with 110 400 in 1956. In addition, the working group on North Pacific whale stocks recommended limitation on catches of all species in this area.

The 16th IWC meeting decided to ban for an extra five years the taking of

Humpback Whales and Blues in the North Atlantic (Iceland and Denmark having agreed to observe the latter ban). However, when the Schedule was amended to prohibit totally the taking of Blue Whales from the Atlantic, objections received from Japan, UK, Norway and the USSR rendered this restriction no more than theoretical.

It was hoped that at this meeting, the IWC would abolish the Blue Whale unit, or at least place tight restrictions on the total catch; the IWC decided to do neither. The IWC set no Antarctic catch limit whatsoever for the 1964-65 season. The reason was that the USSR had become increasingly incensed by the Japanese policy of buying up British, Dutch and Norwegian factory ships, and with them those countries' national quotas. This resulted by 1964 in Japan laying claim to 52% of the Antarctic catch while the USSR had to settle for 20%.

Russia therefore demanded a redistribution of national quotas before she would agree even to a voluntary limit on catches, or to the International Observer Scheme. Japan said there would be no redistribution unless the USSR obeyed the voluntary quota which totalled 8 000 BWu. The result was deadlock and no limits were set. In spite of this, the 16th IWC report says "only 6 958 units" were taken.

1965

After the failure of the 1964 IWC to set any limit on Antarctic catches, a series of non-executive meetings of IWC members were held and it was agreed (though this had to be ratified at the 17th IWC in London) that the quota for 1965-66 should be 4 500 BWu and that lower quotas should be set in the two following seasons. The Scientific Committee warned that this limit was too high: there should be individual species quotas and these should not total more than 2 500 BWu. The taking of Blues should be completely banned in the southern hemisphere and for a five year period in the North Pacific. No Humpbacks should be killed in the North Pacific in the 1966 season. Chile and Peru, which were outside the IWC, should be asked to adhere to the minimum length requirements for the capture of Sperm Whales. The Committee also finally recognised that there was a case for closing the former Antarctic sanctuary, concluding after all that this "might save certain species from eventual extinction".

The IWC placed a five year ban on the taking of Blues from the North Pacific and a one year ban on Humpbacks. But an amendment giving protection to Sperm Whales between 40°S and 40°N was objected to by Japan, Norway and USSR, who used the opt-out procedure of Article V. Total protection was given to Blues in the Antarctic.

No progress was made on the International Observer Scheme, the Antarctic sanctuary was kept open to whaling fleets, and the quota (in spite of the recommendations of the scientists) was kept at 4 500 BWu's, with no separate species limits.

The Antarctic pelagic catch was 4 089 BWu, nearly 90% of the whales taken being Sei Whales.

1966

The Scientific Committee in its evidence to the 18th IWC drew on the findings of the Report of a joint IWC/FAO working party. It was recommended that the taking of all Fin Whales in the Antarctic should cease for a number of years, and that the quota (made up of Seis alone) should be 2 000-2 500 BWu. Failing this, the absolute maximum quota should be 3 000-3 500 BWu. Blues should be protected in the whole of the southern hemisphere, Humpbacks in the North Pacific should receive at least another year's protection, and the catch of Fin Whales in that area should be limited to 1 800.

The IWC extended the Humpback ban, made no move on the International Observer Scheme, and put the 1966-67 Antarctic pelagic quota at 3 500 BWu. The catch was 3 511 BWu.

Concern was expressed about the impact of whaling from land stations in the S. Hemisphere and it was noted that of the nations whaling from land stations, only two (UK and South Africa) were members of the IWC.

1967

The Scientific Committee repeated its recommendations of the previous year, adding that if the Antarctic quota was set well below 3 100-3 600 BWu, then there was no objection to the sanctuary (due to have closed "automatically" in 1958) remaining open.

The IWC chose to continue the North Pacific Humpback ban for three years, and to give protection to Blues in the southern hemisphere. However, the Antarctic "sanctuary" was to remain open for the taking of Fin and Sei Whales, the Antarctic pelagic quota for 1967-68 being set at 3 200 BWu. The catch was 2 804 BWu.

It was decided that nothing could be done about the International Observer Scheme, since this did not feature as a specific item on the agenda.

1968

The Scientific Committee put the present sustainable yield in the Atlantic at 5 400 Sei Whales and 5 000 Fins, giving a total quota of 3 400 BWu. As long as the limit set for 1968-69 was well below this figure, it would be alright to leave the sanctuary open. In the North Pacific, the sustainable yield of Sei Whales had declined to 1 000-2 000 and the Committee argued also for the Fin Whale catch to be restricted to a maximum of 1 600.

The 20th IWC put the Antarctic quota at 3 200 BWu and kept the former sanctuary open. Norway chose not to take up her quota for the 1968-69 season and the catch therefore totalled 2 469 BWu. However, 5 700 Sei Whales were taken − 370 in excess of the estimated sustainable yield. The North Pacific whaling nations agreed at a separate meeting to restrict their catches of Fin and Sei Whales. The International Observer Scheme was added to the Schedule, but was not implemented. Nations were urged to implement the scheme as soon as possible.

1969

Again the Scientific Committee argued for the abolition of the Blue Whale unit and put the Antarctic sustainable yields at 3 300 Fin Whales (although Japanese scientists said the figure should be 4 900-5 600) and 5 000 Seis. The sustainable yield of North Pacific Seis was 1 700 and for Fin Whales 1 640.

At the 21st IWC meeting, nations were again urged to implement the International Observer Scheme, and the North Pacific whaling nations said they would cooperate. The IWC decided to retain the Blue Whale unit rather than introduce separate species limits − they said they would "keep a watchful eye on the situation" − and set the 1969-70 Antarctic quota at 2 700 BWu. The catch was 2 477 BWu, and again included large numbers (5 856) of Sei Whales.

1970

In view of the previous seasons' catches, the sustainable yield for Antarctic Fin Whales was put at 2 700 (although Japan disagreed) and the Scientific Committee put the sustainable yields in the North Pacific at 1 300 Sperm and 3 100 Sei Whales.

The 22nd IWC meeting retained the Blue Whale unit in the Antarctic (the quota for 1970-71 being 2 700 BWu), but set individual species limits for the North Pacific at 1 308 Fin Whales and 4 710 Seis. The IWC found it impossible to set limits on the taking of Sperm Whales, which had never received this protection in either the North Pacific or the Antarctic.

The 1970-71 Antarctic catch was 2 470 BWu, and 6 153 Sei Whales were taken. In the North Pacific, the quotas for Fins and Seis were marginally exceeded and 11 273 Sperm Whales were taken.

There was another call to implement the International Observer Scheme "as soon as possible".

1971

The Scientific Committee, reporting to the 23rd IWC in Washington reiterated their plea for the abolition of the Blue Whale unit. The story was much the same as it always had been. Specific quotas were proposed for Fin Whales and Seis in the Antarctic, and for Sperm, Fin and Sei Whales in the North Pacific.

The Report of the IWC records decisions similar to those offered before. The Blue Whale unit was retained for 1971-72, but it was proposed that it should be replaced by specific quotas for the 1972-73 season. The quota which was set again exceeded the recommended one of 1 933 BWu, being put at 2 300 BWu. The International Observer Scheme was not implemented.

1972

The 24th meeting of the IWC took place two weeks after the UN Stockholm Conference on the Human Environment where the 10 year moratorium on all whaling was first proposed. The moratorium was proposed again at the IWC meeting and was the first test case to see if any of the UN proposals would be implemented. The moratorium was rejected by the Commission on the grounds that "regulation by species and stocks was the only practical method of whale conservation". However, some attempts at long overdue improvements were finally made.

The International Observer Scheme which was first proposed by the Norwegians in 1955 was introduced. "The Antarctic whaling countries signed an agreement to provide for an exchange of observers on each others' factory ships.

Table 1

Voting Records on Moratorium at 1972 and 1973 IWC Meetings:

Country	1972			1973		
	For	Against	Abstained	For	Against	Abstained
Argentina	x			x		
Australia		x		x		
Canada		x		x		
Denmark		x				x
France		x		x		
Iceland		x			x	
Japan		x			x	
Mexico	x			x		
Norway		x			x	
Panama		x		x		
South Africa		x			x	
UK	x			x		
USA	x			x		
USSR		x			x	
Total	4	6	4	8	5	1

N.B. A ¾ majority vote is needed for the Plenary Sessions.

Whaling operations whether from factory ships or land stations would be under scrutiny from an observer, appointed by the Commission, from another nation (so the Russians swapped observers with the Japanese).

The Blue Whale unit was finally abandoned (its abolition was proposed to the Commission by the Special Committee of Three Scientists in 1963) and quotas were set for each species separately (see Table 2). Quotas were also set separately for male and female Sperm Whales.

The Commission also accepted recommendations to:
1)   approach non-IWC whaling nations and ask them to join the Commission.
2)   set up a committee to increase international research efforts. This was rather pompously named the "International Decade of Cetacean Research".
3)   expand the staff and budget of the IWC.

1973

The Scientific Committee stated that "at the present time there is no biological requirement for the imposition of a blanket moratorium on all commercial whaling". However, the proposal for a ten year moratorium was passed by the Technical Committee by a simple majority vote, but failed to get the necessary ¾ majority in the Plenary Session (8 votes to 5, with 1 abstention).

The overall quota was reduced by 1 100. Sperm Whale quotas were divided into three areas in the southern hemisphere to give greater protection to separate populations. The Fin Whale quota in the Antarctic was reduced to 1 450 with the proviso that Fin whaling in the Antarctic should be phased out over a three year period.

Discussions on the expanded Secretariat and the International Decade of Cetacean Research were deferred until the next meeting.

Under the 90 day rule, USSR and Japan objected to the regional divisions for Sperm Whales in the southern hemisphere and to the Minke Whale quota of 5 000 in the Antarctic. Japan also objected to the principle of phasing out Fin whaling in the Antarctic over a three year period.

1974

Brazil re-joined the Commission, bringing the total number of member countries to 15. The proposal for the 10 year moratorium was again introduced by the US delegation. It was not voted on because the Australian amendment to the moratorium — a compromise was adopted by the Commission. The amended decision was only to have effect from the 1975/6 season. It set up a mechanism within the IWC whereby any species considered by the Scientific Committee to be below maximum sustainable yield (msy) level (see pages 68/71) would automatically be put on the protected list until such time as it had recovered to at least that level. This should have meant protection for the Fin Whale in all areas in 1975, and partial protection for the Sei Whale. It would have conferred protection on the Fin Whale one year earlier than had been decided at the 1973 IWC meeting.

The overall quota for the 1974/75 season was reduced by a mere 200 units:

Fin Whales	1 300	(700 less than the 1973 quota)					
Sei/Brydes	6 000	(1 500	”	”	”	”	”)
Minke Whales	7 000	(2 000 more	”	”	”		”)
Sperm Whales	23 000	(same quota as 1973)					

Quotas for the Antarctic were divided into three areas for the first time. This is aimed to stop the complete decimation of separate populations. Area divisions were also set for Sperm Whales in the southern hemisphere again.

The Commission decided to strengthen the Secretariat by providing more money and increased staff to coordinate research activities. It was again agreed that non-IWC whaling nations should be encouraged to join the Commission.

There was no information on the International Decade of Cetacean Research, but the Commission agreed that further coordinated research was necessary.

The FAO presented an interim report on the activities of the FAO/ACMRR* Working Party on Marine Mammals which was critically examining the advisory statements made in recent years by the Scientific Committee. A full report was promised at the next IWC meeting.

The Mexican delegate made a final statement to the effect that if the 10 year moratorium was not introduced soon, "this Commission will be known to history as a small body of men who failed to act responsibly in the terms of a very large commitment to the world and who protected the interests of a few whalers and not the future of thousands of whales".

1975

The long discussion regarding how to put the Australian amendment into practice was finally resolved. Whale stocks were divided into three categories — Protection stocks, Sustained Management stocks and Initial Management stocks — depending on whether they were above, at, or below their maximum sustainable yield level (for discussion of this concept see page 68). The Scientific Committee divided the Southern Hemisphere whale stocks into 6 areas for Minke, Sei and Fin Whales and 9 areas for Sperm Whales, however, this latter figure was reduced to 6 by Plenary Session. The total quota was reduced by just over 5 000 whales. Regarding the errors and uncertainties inherent in all these calculations, the Scientific Committee recorded in their 1975 report "The Scientific Committee

---

Advisory Committee on Marine Resources Research

is faced with an array of possible values which cannot be subjected to proper statistical analysis".

The usual system of allowances operated, which allows the catches for certain stocks to be exceeded by a given amount, provided that the total catch for the Southern Hemisphere for each species does not exceed the total quota.

In the North Pacific, Fin and Sei Whales received protection; Bryde's Whales were given a quota of 1 363; male Sperm Whales a quota of 5 200 and female Sperm Whales a quota of 3 100.

In the North Atlantic quotas were set for the first time for Fin and Minke Whales. The Fin Whale quotas would not be taken up by Canada, but Iceland would take 275. The total quota of 2 550 Minke Whales covers two areas.

The minimum size limit for Sperm Whales was not changed nor was the maximum sustainable yield by weight calculated. No progress was made on the proposals for the International Decade of Cetacean Research.

I. Rindahl, Chairman of the Commission, resigned and was replaced by Mr. A. G. Bollen of Australia. The dominant attitude of the Commission was illustrated by Mr. Rindahl's remark to the Plenary Session: "If we should submit ourselves completely to the Scientific Committee it might be too much of a good thing."

1976

New Zealand rejoined the Commission, bringing the numbers to 16.

In spite of bitter debates over the Scientific Committee's recommended cuts in quotas, the Commission did finally accept all but one recommendation. Although Russia and Japan maintained formidable opposition to reduced Sperm Whale quotas, these were finally reduced more than any others, the total being 6 364 less than for 1975. Had the Scientific Committee's recommended quotas for msy by weight been accepted — the only logical way to manage yield for the Sperm Whale — the figure would have been lower still. A closed season for Sperm Whales was agreed upon to prevent interference with breeding. The usual system of allowances was continued.

The Minke Whale quota was raised substantially as a result of a re-calculation of the initial populations, although the Scientific Committee was unhappy about the available information on the species. Similarly, there was concern over the Sei Whale populations, and a special scientific meeting on Sei Whales was arranged for April 1977. The msy concept on which the IWC's entire management procedure is based had been criticised at the General Assembly of the IUCN* in September

---

*International Union for Conservation of Nature and Natural Resources

1975 (Resolution 8); despite this, this resolution and its proposals for alternative management of whales were not on the agenda for the 1976 IWC meeting.

Further reports on the improvement of killing methods were received and it was recommended that the Secretary should:

1. Contact health authorities in the UK, USA, Japan and other member countries to find out about chemical methods for slaughter and subsequent processing;

2. Check up on methods of killing Minke and other small whales;

3. Find out about US experiments using carbon dioxide, and

4. Ask member nations whether they were considering research into high velocity projectiles to decrease death times.

A motion restricting trade in whale products to member countries was rejected, reflecting Japan's interest in continuing to establish more whaling stations outside the IWC quota system, and Russia's interest in maintaining trade in sperm oil with non-member countries.

It was suggested that the IWC should control small cetacean catches, but the Japanese pointed out that this would have to be referred to the Working Group which is currently looking at redrafting the Convention. The report of the small cetacean sub-group was mainly concerned with animals taken for their own value — the Striped Dolphin, Dall's Porpoise, Harbour Porpoise and the Bottlenose Whale, a species regarded as having been severely over-fished by Norway. It also recommended new management procedures for stocks about which very little is known and better reporting of small cetacean catches.

Spain and Portugal were again urged to join the IWC, as they catch whales from stocks killed elsewhere by IWC countries.

Dr. Ray Gambell (UK) was appointed Permanent Secretary to the Commission, which has established its offices at Histon, Cambridge, England.

1977

At the 29th Meeting in Canberra, Australia, Netherlands rejoined, bringing the membership to 17. Quotas were dropped by an unprecedented 36%. Msy by weight was not seriously considered (had it been, 558 less whales would have been killed). Even so, huge cuts in quotas for North Pacific Sperm Whales were recommended, giving total protection to the males. Simultaneously, Minke Whale quotas in the Southern Hemisphere were cut by over a third and Sei Whale quotas by over one half. In considering Antarctic species the meeting had to consider the interdependence of whale and seal species on krill, and the changes occurring made the application of the msy principle more difficult than ever. The question of risk in the present management procedure and also the IUCN resolution were

discussed, though briefly, and no concrete recommendations emerged.
A Resolution urging governments to stop importing whale products from non-member nations was passed (in a weaker form than originally proposed), as was another Resolution to control the transfer of whaling equipment. Observers from accredited non-governmental organisations were to be admitted to Scientific Committee meetings if approved by its Chairman and Vice Chairman and the Secretary to the Commission.

A review of the North Pacific Sperm Whale quotas was arranged for November 1977, to be followed by a Special Meeting of the Commission (in Tokyo) in December to agree any changes. Both meetings precede the North Pacific whaling season. A special scientific meeting on Minke Whales was agreed for 1978.

During the year Japan issued itself a Scientific Permit to take 240 Bryde's Whales from a protected stock, but surprisingly little information was available from the catch. An administrative agreement was made that, before Scientific Permits for catches are issued, the Scientific Committee would be asked to comment on both the permits and the programmes of which they are a part.

The Antarctic Whaling Season was abolished.

The Scientific Committee recommended a cessation of whaling for Bowhead Whales in Alaska.

The revision of the International Whaling Convention was to be discussed at a Preparatory Meeting in Denmark in the Spring of 1977.

Table 3

IWC 1974

Commissioners representing:	Argentina	Australia	Brazil	Canada	Denmark	France	Iceland	Japan	Mexico	Norway	Panama	S.Africa	USSR	UK	USA
No. of advisers:	—	5	2	6	1	1	2	18	1	6	—	3	5	4	16

Scientific Committee composed of representatives from:

Australia	(2)
Canada	(2)
Denmark	(1)
FAO	(2)
France	(1)
Iceland	(1)
Japan	(4)
Norway	(1)
S. Africa	(1)
USSR	(3)
UK	(3)
USA	(4)

Technical Committee composed of all Commissioners

Infraction Sub-Committee composed of a representative from:

USSR
Japan
Denmark
S. Africa

Administration & Finance Committee composed of delegations from:

Canada
Iceland
Japan
USA
USSR

Non-Governmental Observers: Fauna Preservation Society, International Union for the Conservation of Nature & Natural Resources, World Federation for the Protection of Animals, Friends of the Earth Inc., Friends of the Earth Ltd., World Wildlife Fund, Project Jonah, International Commission for S.E. Atlantic Fisheries, United Nations Environment Programme, Sierra Club, Whaling & Marine Archives, International Society for the Protection of Animals, Food & Agriculture Organisation.

Table 2 – QUOTAS AND CATCH FIGURES FOR IWC COUNTRIES BY AREA

Area	1971-1972 quota	1971-1972 catch	1972-1973 quota	1972-1973 catch	1973-1974 quota	1973-1974 catch
S. HEMISPHERE						
Fin		2,734a	1,950*	1,803g	1,450	1,311m
Sei	2,300	5,619b	5,000*	4,364h	4,500	4,857n
Brydes	BWu*					
Minke		3,157c	5,000*	5,920i	5,000	7,830o
Sperm:		9,928d				
male	2,747x		8,000	6,986j	8,000	6,677
female			5,000	4,330	5,000	4,497p
N. PACIFIC						
Fin	1,046	758	650	460	550	413
Sei	3,768	2,528	3,000	2,585	3,000	2,643
Brydes						
Minke						
Sperm:	10,841	6,323				
male			6,000	4,606	6,000	4,419
female			4,000	3,962	4,000	3,708
N. ATLANTIC						
Fin		599e		269k		289q
Sei		315		139		9
Brydes						
Minke		97				
Sperm		120f		47l		71r
TOTAL		32,178	38,600	35,475	37,500	36,724

a Not including 9 Peru, 1 Brazil
b Inclusive of 3 Brydes
c Not including 702 Brazil
d Not including 66 Brazil, 1,550 Peru
e Not including 90 Spain
f Not including 327 Azores, 63 Madeira, 181 Spain
g Not including 11 Peru
h Not including 330 Peru, 7 Brazil

1974-1975		1975-1976		1976-1977		1977-1978	
quota	catch	quota	catch	quota	catch	quota	catch
1,000	1,001s	220	206	0	0	0	
4,000 (3 areas)	4,017t	2,230 (6 areas)	1,820	1,863 (6 areas)	1,861	771	
		0		0	225	0	
7,000	8,149	6,810 (6 areas)	6,034	8,900 (6 areas)	8,676	5,690	
8,000 (3 areas)	7,097u	5,870 (6 areas)	4,022z	3,894 (9 areas)	3,308	4,538	
5,000 (3 areas)	4,737	4,870	3,024	897 (9 areas)	767	1,370	
300	162	0	162	0		0	
2,000	1,941	0	508	0		0	
		1,363	1,433	1,000	1,340	524	
			370	541 (3 areas)	360	400	
					7,211		
6,000	4,261	5,200	4,261	4,320	4,200	0	
4,000	3,598	3,100	3,610	2,880	3,011	763	
	245	363 (4 areas)	246	344 (5 areas)	275	459	
	138		141	132 (2 areas)	3	84	
					0	0	
		2,550 (2 areas)	2,140	2,483 (4 areas)	2,343	2,555	
	37	363	37	344	111	685	
37,300	35,383	32,939	32,640	27,939	27,484	17,839	

i   Not including 650 Brazil
j   Not including 75 Brazil, 1,497 Peru
k   Not including 73 Spain
l   Not including 311 Azores, 77 Madeira, 178 Spain
m   Not including 5 Peru
n   Not including 3 Brazil, 174 Peru
o   Not including 765 Brazil
p   Not including 29 Brazil, 1,286 Peru

q   Not including 57 Spain
r   Not including 167 Spain
s   Not including 5 Peru
t   Not including 545 Peru
u   Not including 793 Peru
*   Antarctic only
x   20° - 70° East only
z   incomplete

CHAPTER 5

WHALING NATIONS

Not all the countries which belong to the IWC hunt whales, and conversely, not all the whaling countries belong to the IWC.

Chile and Peru are important whaling countries which belong to another group (which includes Ecuador): the Permanent Commission for the Exploitation and Conservation of the Marine Resources of the South Pacific, which was established in 1952. This division of whaling nations into two groups makes the regulations of both inherently weak. An example of this weakness was the 1967 ban on Blue Whale hunting by IWC countries. In the same year, Japan avoided the restrictions by allowing its whalers to form companies in Chile and then continuing Blue whaling under Chilean flags of convenience. (To our knowledge, Japan has part or wholly owned whaling companies in Chile, Peru, Brazil and probably South Korea. There may be more.)

The rules of the Permanent Commission state that Grey and Right Whales can only be caught when they are for human consumption, Blue and Humpback Whales can only be caught when over 21.30 and 10.60 metres long respectively, and when annual quotas are set they are in Blue Whale units. These rules were made in 1952 but they still (1978) apply.

According to a paper published by the Peruvian Ministry of Fisheries in 1973, "Comparative Study of the Economic Importance of Marine Resources in the 12 and 200 Mile Zone", the Peruvian whales consist of the following species: *Kogia Breviceps* (Pygmy Sperm Whale), *Balaena glacialis* (Southern Right Whale), *Megaptera novaeangliae* (Humpback Whale), *Balaenoptera borealis* (Sei Whale) and *Physeter catodon* (Sperm Whale). Two of these species, the Humpback and the Right Whales, are protected under IWC regulations. The catch consists of Fin, Sei, Brydes and Sperm Whales. The data published by the BIWS (Bureau of International Whaling Statistics) give no account of Right Whale or Pygmy Sperm Whale catches anywhere in the world. It is clearly important that all whaling nations are subject to the same regulations.

Both Chile and Peru send in the reports of their catches to the BIWS, but sporadically and often many years late — presumably to avoid reprisals. At the 1977 IWC meeting the latest available figures were the 1974-5 Peruvian catch and the 1975-6 Chilean catch. The IWC makes very little effort to rectify the situation. During the 1970/71 season 3.1% of the total baleen oil production was produced by non-IWC countries. The percentage of sperm oil produced by non-IWC members (Portugal, Spain, Peru, Chile and the Bahamas according to BIWS) was over twice as much — 6.3%. During the 1971/72 season, the percentage rose to 6.6%. If one considers only non-pelagic whaling nations (ie excludes Japan and USSR), non-IWC countries account for over 30% of all sperm oil production. The significance of non-IWC whalers is thus more apparent. It should also be borne in mind that this percentage refers only to oil yield;

some of the non-IWC countries are less efficient at oil extraction and the percentage by total weight of whales could be higher.

Meat imports are also significant — in the first 5 months of 1977 over 12% of Japan's whale meat imports (2,809 tons) came from non-IWC countries, principally Peru and South Korea. Peruvian whale meat fetched US $730 per ton.

a)   IWC COUNTRIES WHICH WHALE

(Catch figures are taken from the Bureau of International Whaling Statistics, Norway.)

Australia
One whaling company, Cheynes Beach Holdings Ltd, centred on a small landbased station at Frenchman's Bay near Albany in West Australia. It employs 3 chasers and a spotter plane. (Two of the company's directors attend the IWC annual meetings.) Cheynes Beach only takes Sperm Whales. 1,172 were taken in the 1974/75 season.

Recent reports on findings in Australia have shown dramatically high levels of mercury in whale flesh. It would be ironical if the whales might eventually be protected by their high levels of toxic heavy metal. According to the Australian Financial Review, while the National Health and Medical Research Council of Australia has set a critical level for mercury content of sea life at 0.5 parts per million, examinations of Sperm Whales in Western Australia show a mercury level as high as 12.8 parts per million in some parts of the mammal. Australian health and agriculture officials are concerned about the implications of the findings for Australian diet because an increasing amount of whale meal is being used for chicken, pig and cattle feed. Information provided by the Western Australian Commissioner for Public Health indicates high levels of mercury in chickens and eggs as a result of the use of whale meal at chicken hatcheries. (Sierra Club International Report, August 1974, Vol. 11 No. 21)

Brazil
One shore station, Cia. de Pesca Norte do Brazil (partly Japanese owned) and one catcher ship. 1,039 Minke Whales, 54 Sperm and 3 Sei Whales were taken in 1975. Brazil rejoined the Commission in 1974.

Denmark
Greenland: Large scale "aboriginal" whaling. 204 Minke, 2 Killer Whales, 907 White Whales, 169 Narwhals, 596 Common porpoises and 13 others caught in 1974 (see also p. 99). In 1976, Greenlanders killed 5 Humpbacks (under an IWC "aboriginal" quota of 10) and 5 Fin Whales.
Faroe Islands: 673 Pilot Whales were driven into shallow waters and killed in 1974 (see also p. 111).

Iceland
One shore station and 4 catchers. The Fin Whale is the most important species:

245 were caught in the 1974/75 season as well as 138 Sei Whales and 37 Sperm Whales.

Japan (*For a full discussion of Japanese Whaling see pages 46/61.*)

Norway
Norway stopped Antarctic whaling in 1968 and Arctic whaling in 1971.
1,820 Minke Whales, 1 Pilot Whale and 6 Killer Whales were caught in the 1973/74 season.

USSR
The Russian fleets were reduced in number in 1976 from three to two. This is more a reflection of economic necessity than it is of concern for whales. The voting record of the USSR, solidly behind Japan in attempting to relate the quotas for Sperm Whales in the Southern Hemisphere, is a truer indication of where the Russian sympathies lie.

b)   COUNTRIES WHICH WHALE, BUT WHICH DO NOT BELONG TO THE IWC

Chile
1 shore station owned by Japan. In 1975, 62 whales were taken. A new whaling station, reportedly backed with Japanese capital and technical expertise, is under consideration by the Chilean government. It would take 1,500 Sperm Whales in the next three years (1977 report).

China
No data on China's whaling are available. China recently bought a catcher ship from Taiyo Fisheries (Japan). Chinese whale meat is currently on sale in supermarkets in Japan.

North Korea
While there is no data to show that North Korea whales, she exports whale meat to Japan. In the first 5 months of 1977 this amounted to 10 tons only.

Peru*
1 whaling company, Compania Bellerna Kinkai SA (2/3rds of its capital investment is Japanese). 3 catchers. 11 Fin, 330 Sei, 1,497 Sperm Whales were caught in the 1974/75 season. Most of the meat and whale products are exported to Japan. There are no inspectors and even the most basic knowledge of the offshore whale populations is lacking. The Sperm Whales are probably taken from one of their breeding areas and the rules governing length mean that the breeding males are killed. In 1975 the Japanese company was taken to a Peruvian court, but the case against them — of infringing the South Pacific Convention — was dismissed for lack of evidence. The Japanese are hoping that the Peruvian government can be

*Ministeria de Pesqueria. Informe de la Inspeccion a la Cia. Ballenera del Kinkai SA. Undated. "Whaling in Peru", an address by Felipe Benavides, November 11, 1975. Pers. Comm. WWF Lima, Peru. Dr Augusto Urrutia.

persuaded to allow them to hunt Blue Whales again as they maintain they have seen great numbers of them off the coast.

In 1974 an inspection of the Kinkai Co. plant at Paita and their whaling operations was carried out by a Mr. Chang, for the Ministry of Fisheries, During this inspection 26 Sperm Whales were caught of which 24 did not reach the minimum dimensions listed in the South Pacific Regulations. He was told that generally whales are measured by crewmen on the boats and thrown back to sea if they were less than the minimum. Mr. Chang recommended that permanent Government personnel should be maintained during the hunting season to exercise strict control and to substantiate reports submitted by the Kinkai Co; also that the Japanese should submit with their monthly reports, invoices of internal and export sales of different products marketed. Considerable dissatisfaction was voiced by the Peruvian employees and the Japanese whalers were unforthcoming. Although the Sea Institute was asked to make a technical report so that realistic quotas could be set, this has not taken place, and the recommendations for stricter sanctions and penalties seem to have been forgotten.

Portugal
Azores: 311 Sperm Whales were taken from one land station during the 1974 season. Large numbers of porpoises and dolphins are also taken, but not reported. Madeira: no data for recent seasons. In 1973, 77 Sperm Whales were caught. In 1959 1 Right Whale was killed, and the 1977 IWC Scientific Committee recorded a plea to the Portuguese Government that no further Right whales be killed.

South Korea
Large whaling (Fin and formerly Humpback) is conducted from one shore station at Ulsan on the coast. Small whaling (mainly Minke) is conducted from several ports. The South Koreans have been whaling since 1965 at least, according to BIWS figures, and just recently there seems to be a slight shift from "small" whaling to "large" whaling. (1972, 1 Fin and 767 Minke caught; 1973, 4 Fin and 882 Minke; 1974, 52 Fin and 566 Minke.) No data, but we suspect that the Korean whaling companies are wholly or partly owned by the Japanese. In the first 5 months of 1977 South Korea exported 949 tons of whale meat to Japan at a value of US $1,439 per ton.

Spain
167 Sperm Whales, 57 Fin Whales taken during the 1973/74 season. 136 whales, thought to be mostly Fin Whales were killed in 1976. The average catch of Fins from 1970 − 74 was 92 whales.

Sierra Fishing Agency
A pirate floating factory cum catcher ship operating outside IWC control and, in 1975, flying a Somali flag of convenience, is now flying that of Cyprus. The Beacon Sierra agent is now in Limassol, although the owners are reported to be in the Forrentningbanken, Oslo, Norway. According to information available

in 1975, 4 Japanese Nationals are always aboard to represent the Japanese company that buys the meat of the whales caught. From an animal of 25-30 tons, 5-6 tons of prime whale meat is taken, the rest is dumped. The meat is then frozen and labelled, with no apparent justification, "Produce of Spain".*

Sierra's operations are in an area prohibited to factory vessels of IWC member nations for the taking of baleen whales, yet evidence suggests that Sierra's catch is exclusively baleen whales. She is registered inaccurately as a fishing vessel and has allegedly carried a cargo of whale meat into Cape Town, South Africa where this cargo is restricted.

In view of these numerous offences of omission and commission, Sierra's reported catch figures must be regarded as open to question. These figures were, in 1974, 499 Sei Whales and 2 Fin Whales. In the first 5 months of 1977 Somalia exported 586 tons of whale meat to Japan.

*Sierra information from Nick Carter, International Society for the Protection of Animals.

Table 4 – NUMBERS OF WHALES CAUGHT BY SOME NON-IWC COUNTRIES

Years	Chile (mainly Sperm Whales)	Bahamas	Somalia	Peru (mainly Sperm Whales)	Portugal (mainly Sperm Whales)	Spain (mainly Sperm Whales)	South Korea (species unknown)
1960/61	2,334			3,602	507	292	
1961/62	2,338			3,301	583	198	
1962/63	1,543			3,241	658	126	
1963/64	1,508			2,066	611	219	
1964/65	1,348			1,289	530	280	266
1965/66	1,099			1,365	410	283	323
1966/67	744			645	395	287	356
1967/68	*			2,462	149	359	344
1968/69	253	446		2,305	228	315	421
1969/70	291	628		1,935	249	413	740
1970/71	253	602		1,773	353	361	755
1971/72		72	76	1,896	390	271	769
1972/73			491	1,838	388	251	886
1973/74			451	1,812		224	618
1974/75	106		276	1,343	150	*	574
1975/76	62					136[1]	533#

*No information supplied

(1) Said to be Fin Whales

#To October 1976

CHAPTER 6

THE ECONOMICS OF WHALING

INTRODUCTION

The most telling argument in favour of whaling has always been the simple one —
that it was profitable to do so. All other arguments are subsidiary to this one and,
although the Japanese whaling industry has recently been making much play with
claims about traditional eating habits, protein needs, employment prospect and so
on, one cannot but doubt that they would continue to put such points so strongly
were whaling to become unprofitable. In this situation it is worth reviewing the
economics of whaling both as a means of anticipating possible responses by the
industry to falling whale populations and other pressures, but also of determining
what might be the best future for whaling from a world community, rather than a
private industry, point of view.

In this section we will be concentrating primarily on the Japanese whaling industry
for two main reasons:

1)   Because the Russians have released little data on their industry, and

2)   Because the Japanese Whaling Association (JWA) have waged a strenuous
      propaganda campaign which has often used erroneous or misleading
      arguments which we wish to correct.[1]

We feel that the omission of data on the Russian whaling industry does not
significantly alter the strength of the arguments we put, since although there might
be different priorities as to who should benefit from the exploitation of the
resource, there is agreement as to the value of doing so. Thus the economic
arguments that apply to the Japanese industry would also apply to the Russian
industry. It is unlikely that slaughtered whales would be interested in the nicety
of distinction between state capital and private capital.

a)   THE CURRENT POSITION

The Importance of Whale Meat in Japan:

Consumption
Whale meat constitutes a significant, but not an indispensable part of the Japanese

---

(1) One publication, "The Position of Whaling in the Japanese Economy", by the Economist
Intelligence Unit (EIU) (1974), was commissioned by the JWA "to work out a report on
whaling and its importance to the Japanese economy. The report in part supports the
Japanese point of view but also provides enough ammunition for the Japanese Government
to justify itself to public opinion at home should it be forced to do so **after bowing to world
pressure to call a halt on whale hunting**". (Asahi Evening News, June 24, 1974) It may be
observed that the EIU report as published does not reveal the fact that it was commissioned
by the JWA.

diet. In their booklet, "The Importance of Whaling", the Japan Whaling Association makes two very misleading assumptions:

The first is that as the "whalers are catching whales on the basis of 'scientific research' there is no danger to the whales". This is no justification if the results of the scientific research are in dispute (see p. 62). and in any case the Japanese often disregard the scientific advice of the IWC when it conflicts with their commercial interests (see p. 32).

The second is that Japan's whaling activities are **indispensable** for maintaining an adequate level of animal protein for its people. This is not true. Whale meat is a cheap form of animal protein in Japan which is not indispensable as substitutes do exist. In another more recent booklet, "Whaling — an Important Part of Japan's Food Industry", the Japan Whaling Association goes further: "For the forseeable future a significant proportion of the supply of protein for the Japanese people can **only** come from whales". This so-called significant proportion was, in 1975, 0.8% of total protein, and has been declining. The official Japanese statistics show that both the catch and consumption are decreasing. In the period between 1965 and 1973, consumption had dropped from 203 000 to 122 700 metric tons.* As a result of this "rising meat production and consumption whale meat has come to occupy a smaller share of indigenous meat production in recent years. In 1970, home whale meat production, at 139,000 metric tons, accounted for about 9.9% of total home meat output; in 1972 the equivalent figure was 7.5%." (EIU report) In early 1974 Japan also exported whale meat to the value of 390,000 yen (approximately £570.00) to Indonesia.

Table 5a — International Comparisons of net Protein intake — 1970
(per head per day)

	% Meat	% Fish & Shellfish	% Milk	% Non Animals	Total Animal Products
Japan	8.7	20.8	5.2	58.6	41.4%
UK	27.4	4.6	23.8	38.5	61.5%
USSR	15.1	3.3	16.6	56.4	38.8%
USA	38.3	4.0	24.5	27.5	72.5%

NB — In all tables per capita protein requirements shown are in excess of those recommended by the WHO (World Health Organisation).

Whale meat, says the JWA, is especially important to the Japanese as it is a cheap source of protein. It quotes as an example that 230,000 tons of beef would have to be imported to replace whale meat and that this would cost the balance of payments about $500 million a year.

The most recent figures indicate that only 65,000 metric tonnes were consumed in 1975, less than half the 1974 figures. This represents a total drop of 68% over a ten-year period.

There are two main arguments against this kind of reasoning: first, it is ridiculous to give an example of the most expensive form of animal protein when trying to find a substitute for one of the cheapest forms of protein. It would be much fairer to compare prices with a similarly priced protein such as mutton. Mutton is available from Australia and New Zealand; it is unpopular but cheap (85 yen for 100 grams, compared to 100 yen for 100 grams of whale meat).

Second, in any event, forced alternatives rarely fail to cost something extra, as, by killing whales, the Japanese are exploiting a 'free' resource. It was equally 'cheap', for example, to eat the Dodo.

Table 5b – Consumption of animal protein in Japanese city households
(annual average in kilograms)

	Fresh/ frozen fish	Pro- cessed fish	Beef	Pork	Pro- cessed pig	Chicken	Whale	Total	Whale as % of total animal protein	Whale as % of non- fish protein
1965	43.4	9.6	9.6	9.8	5.8	5.2	2.9	86.3	3.3%	8.7%
1970	36.6	9.1	7.5	14.7	7.5	9.5	1.5	85.9	1.7%	3.7%

Source: Fishery Statistics of Japan 1970

Traditionally, the Japanese have been fish eaters. The JWA claim that whale meat plays an important part of their animal protein intake and seem anxious that the Japanese should follow western countries' eating habits by consuming a large proportion of animal protein other than fish. Before the Meija Era (1867-1912), the Japanese did not eat any animal flesh as they were prevented from doing so by their religion. But, as the Buddhist priests had conveniently labelled whales as fish, their meat could be eaten. The Japanese only started to eat pork, beef, etc about a hundred years ago. Perhaps they should seriously consider whether it is wise in terms of protein and cost to copy the eating habits of the west, especially at a time when the west, because of rising costs, etc., will have to revert to eating less meat.[1]

---

(1) cf. Friends of the Earth publications: Losing Ground (1974) and World Food Crisis (1974).

Table 6 – Japanese Food Consumption (from UN data)
(Grams per capita per day)

	Cereals	Potatoes	Pulses	Meat	Milk	% Animal protein	Calories
1934-38	432	127	46	8	9	–	–
1960-62	411	181	46	21	69	9	2 330
1963-65	404	181	45	28	96	11	2 410
1966-68	379	174	46	36		13	2 450
1969	360	162	47	41		14	2 450
1970	350	161	44	48		15	2 470

In the FAO Agricultural projection 1970-80, Japanese meat consumption was projected as rising to 24.2 kgs. per year per capita from 15.7 kgs. (1970)

Substitutes

Although the Japanese claim to be working hard to produce substitutes for whale meat, it is by no means clear that substitutes are really necessary. For example, in 1970, the weight of fish and related products exported for food by Japan was five times that of her whale catch for food, even recognising that much of this is high value produce such as tuna. To solve their food shortage, only a minor cut-back in the export of fish would be needed. However, according to the EIU report on "The Position of Whaling in the Japanese Economy", it is thought that protein from . . . substitutes would probably be accepted in Japan should whaling be prohibited". In fact, alternative proteins are already being produced. Several companies have collectively been producing about 1 000 tons of artificial protein powder a month from soya beans and production has been showing an annual increase of about 50%. With this powder, about 500 tons per month of artificial meat is currently being produced. Again according to the EIU report, "the market for artificial meat seems to be promising and many manufacturers have been trying to sell it to schools, hamburger stands, etc". (Other manufacturers have plans for obtaining protein from plankton in the Antarctic Ocean.)

Production

The Japan Whaling Association has constantly sought to show both at home and abroad[1] that the output and value of the whaling industry is increasingly important. This is just not so. The consumption of whale meat dropped from 203 000 tons in 1965 to 138 000 tons in 1970 (Japanese Fishery Statistics 1970) and down to 65,000 metric tonnes in 1975. Table 7 below shows that although

---

(1) Until recently, the JWA has spent most of its time trying to convince the British and Americans that whaling is a vital part of their way of life. However, during the IWC week in 1974, the JWA and Japan Fisheries Association felt it necessary to run opinion ads in leading papers in Japan explaining why Japan must catch whales and seeking opinions on whether whaling should be banned to conserve whales. (Japan Times, June 28, 1974)

the volume of whale product output has decreased, the value has increased (this makes it profitable for the whalers to continue whaling), despite the fact that whales are getting scarcer. Inflation is also a factor in the price increase; the rate of inflation of food prices is significantly higher in Japan than most other countries.

Table 7 – Volume and value of whale products output
(quantity in metric tons)

	Whale oil	Meat	Others*	Total	Value (Yen '000)
1968	81 213	156 766	16 850	254 829	23 670 000
1969	72 467	136 024	15 443	233 934	22 110 000
1970	72 604	139 230	15 836	227 670	27 190 000
1971	71 775	135 009	16 319	223 103	31 380 000
1972	61 196	121 926	11 350	194 472	27 770 000
1973	50 250	97 921	15 794	163 965	27 320 000

*includes edible by-products other than meat – meal, extracts, solutions, etc.
Source: Ministry of Agriculture & Forestry, Japan

Whaling is only a small part of the total Japanese fishing industry. In 1965, whales constituted over 3% by weight of the output of the Japanese fishing industry, but this had fallen to just over 1½% weight by 1970. To get a real perspective of the comparative unimportance of the present whaling industry, one has to realise that the seaweed culture industry was worth about 3 times as much as whaling in 1970.

In 1965 whales were 6.5% of the total value of Japanese fisheries, but by 1972 this had dropped to a mere 2.5%, *and in 1975 the figure dropped to less than 1%.*

Table 8 – Whaling as a Percentage of the Fishing Industry
(output value in million yen)

	Fishing Industry	Whaling Industry	Share of Whaling in Fishing Industry
1968	768 420	23 670	3.0%
1969	870 870	22 110	2.5%
1970	996 350	27 190	2.7%
1971	1 120 800	31 800	2.7%
1972	1 206 000	27 770	2.5%

Source: Ministry of Agriculture & Forestry, Japan

Figure 1 — Japanese Fishery catch 1962-71

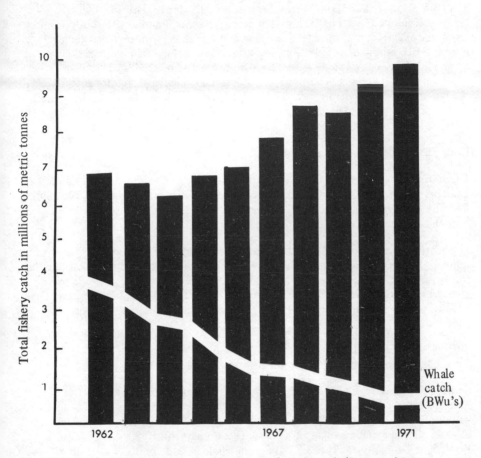

The EIU* claimed that 12,000 people were employed in whaling yet the International Statistics published in Norway gave a figure of only 9,676 in 1963/64 when the industry was at its peak. Only 6,000 people were **directly** employed by the seven large whaling companies in Japan and 200 people were also employed by the ten family-owned companies. With the advent in 1976 of the new Japan Joint Whaling Company (see p. 54), half of the 3,000 employees of the old companies will find themselves the unemployed recipients of $250,000 compensation from the government. It is also doubtful whether the closing of the whaling industry would result in mass unemployment as the JWA claim, as in the Japan Government's White Paper on Fisheries (1972) it is noted that fishermen were in short supply. (There has been a general trend

*in 1974

away from primary industry to white and blue collared jobs in factories in the towns, which has affected the fishing industry.)

The JWA also say that 300 000 people are indirectly dependent on whales — "mainly employed in shops". This is, of course, a weak argument. Nobody would argue, for example, that butchers' shops would close if veal were not available. Even the EIU admit that "relatively few establishments would close and their employment be lost if the retailers ceased to sell whale products". It is also argued in the EIU report — without any published evidence — that local communities might collapse if whaling ceased. If there is a real possibility of this happening, this is the type of problem that should be openly discussed at the IWC since, in any event, the extinction of whales will have the same long-term result.

Uses of Whale Meat in Japan

The Japanese have a long tradition of eating the red meat from whales. Red meat ('aku-niku') is taken from the backs of whales and is generally frozen before being sold on the market. The tail meat ('ono-mi') is considered the most palatable and is often eaten as raw meat ('sashimi').

The ventral meat ('sunoko') and the chest meat ('munaita') are generally used for canning. Roasted (flaked and unflaked) whale meat appears to be the most popular form of canned whale product in Japan. This involves the boiling of whale meat in a sugar and ginger seasoning prior to roasting. Boiled and sliced whale meat (also seasoned in sugar and ginger) is also popular. Sometimes vegetables are added for variety.

The tail of the whale is either eaten raw or is salted ('oba') and is also considered a great delicacy. Blubber and belly meat ('unesu') are also salted.

Some whale meat is used in the production of fish sausage and ham; some is smoked. What is not consumed as food is generally converted into feed for animals. About 10% of the total whale meat is utilized in this form. (Taken from Tanikawa, Eiichi, "Marine Products in Japan — Size, Technology and Research", Koseisha-Koseikaku Company, Tokyo, 1971.) Sperm whale meat from the tail, blubber and cartilage is a minor part of the Japanese whale meat diet.

The Structure of the Japanese Whaling Industry until 1976

The Japanese whaling industry was dominated by the three largest fishery companies in Japan — Taiyo Gyogo, Nippon Suisan and Kyokuyo.

Table 9 — Major Fishing Companies engaged in Whaling, Japan

	Capital (Yen million)
Taiyo Gyogyo KK	15 000
Nippon Suisan KK	10 000
KK Kyokuyo	5 300
Nippon Hogei KK	360
Nitto Hogei KK	100
Hokuyo Hogei	50
Sanyo Hogei	2

Source: Japan Whaling Association

Only the first three companies sent whaling fleets to the Antarctic. Most of the other companies whaled in the North Pacific. NB the capital relates to the total capital of the entire company, not to the capital relating to their whaling activities.

Taiyo was the largest of the three companies, employing a total of 11,069 people, only a fraction of whom were employed in whaling. At the 1974 IWC meeting Taiyo disclosed that they had decided to phase out whaling. It has already sold one of its catcher ships to China. Nippon Suisan, the second largest company, employed 7,067 people. Kyokuyo employed 3,193 people. 12% of its sales related to whaling. In November 1973, the Kyokuyo Company converted Kyokuyo Maru 2 — a 16,400 ton whaling factory (mother) ship — into a waste oil processing vessel which it planned to put into operation in Osaka Bay and the Inland Sea during the spring of 1974.

In addition to the seven large whaling companies listed in table 9 above, there were ten small companies which whaled along the Japanese coast. They owned ten ships which were mostly under 50 DWT (dead weight tons). The seven major companies belonged to the Japan Whaling Association. The ten smaller companies had their own organisation. The Japanese Government issued a whaling licence to each whaling firm within the framework of the quotas allocated to Japan by the IWC. The Government had its own inspectors aboard each factory ship and inspected each land station to enforce whaling regulations.

Table 9 — Composition of the Whaling Industry

1. Whaling ships:

Factory ships	3
Catcher boats	45
Refrigeration ships, carriers, etc.	36
Small catcher boats owned by local companies	10

2. Whaling bases:

Hokkaido Island	4

Honshu	3
Kyushu	2

3. Processing factories owned by large companies:

Taiyo Gyogyo KK	4
Nippon Suisan KK	1
Nitto Hogei KK	3
Nippon Hogei KK	3
Sanyo Hogei Co.	(use one processing factory owned by Nippon Hogei KK)

Source: The Position of Whaling in the Japanese Economy, EIU report *1974*

(This includes Japanese controlled operations elsewhere — see page *42*.)

On June 22nd, 1974, the Tokyo Maritime Safety Department revealed the fact that a modern whaling factory ship, the Miwa Maru, had poached 23 small whales 6-9 metres in length, 180-360 kilometres southeast of Cape Nojimazaki, Chiba Prefecture. The exposure of this offence came at a particularly sensitive time, as Yoshihide Uchimura, Director-General of the Fisheries Agency was quick to point out, "It is most deplorable that such a case should come to light just when the 26th meeting of the IWC is opening June 24. We are worried that the case will be reported overseas as evidence of inadequate supervision by government officers concerned with whaling companies engaged in unethical practices". Yoshihide Uchimura's announcement reflects a growing sensitivity on the part of the Japanese to anti-whaling factions abroad.

The Japan Whaling Association reports that at the end of 1973 Japan had 84 ships engaged in whaling. These include 3 factory ships ranging from 10 000 to 20 000 DWT, several cold storage ships of about 10 000 DWT and 48 catcher ships equipped with harpoon guns, ranging from 300 to 1 000 DWT. In addition, there were 10 catchers of less than 50 DWT engaged in coastal whaling.

Japanese Whaling 1976

The Japanese whaling effort was reduced in 1976 to 2 fleets (2 factories with 43 catchers and 12 catchers from Japanese North Pacific shore stations). In February 1976 the six leading Japanese fishery companies closed down their separate whaling operations and formed the Japan Joint Whaling Company. The government is understood to be giving financial support in the form of loans extended through the Japan Development Bank for purchase of equipment and boats from the six companies. They will also provide $1 million for scientific research previously carried out by the companies and are allowing only $250,000 compensation for half of the 3,000 employees of the 6 companies who will find themselves unemployed.*

*This information is taken from two articles: "Japan Whaling Industry Struggles for Survival" Japan Times, March 3, 1976, and "Japanese whalers in deep water" Far Eastern Economic Review, March 26, 1976.

With the drastic quota cuts for sperm whales in the North Pacific, the Russians and Japanese are considering amalgamating their effort in that area, by mounting a single, multi-national expedition with one factory ship and a number of catchers.

Whaling as a Percentage of GNP

The total GNP in 1973 was 114,227,000 million yen. The whaling industry's share was 27,320 million yen − 0.024%. Preceding years are shown below:

Table 10

1968	0.046%
1969	0.037%
1970	0.038%
1971	0.040%
1972	0.031%
1973	0.024%

As Table 10 above shows, the contribution of the whaling industry to the GNP is small and declining.

Diversification

Japanese fishery companies, realising that whaling is a declining industry, have been diversifying. They have not only continued to develop themselves into general food manufacturing companies, through the production of frozen foods other than frozen fish, but they have also established 77 joint ventures in 38 countries and are importing the products of those newly-formed companies. This investment amounts to 7 717 million yen (March 1972, Japan Fishery Association) of which 242 million yen relate to whaling. There are no indications that this side of the industry will not continue to grow rapidly and thus be able to absorb any redundancy created by the running down of whaling.

One significance of these joint ventures is that Japanese companies can continue whaling operations from countries which do not belong to the IWC (see Chapter 5).

Another aspect of diversification came to light after the 1974 IWC meeting. The Japan Times (June 29, 1974) reported that two whaling towns, Oshika in the Miyagi Prefecture and Taiji in the Wakayama Prefecture, are considering putting much more emphasis on tourism as opposed to whaling, owing to "growing pressure from conservationists abroad for stricter controls on whale hunting".

Land Stations

There are four land stations in the northern-most island of Hokkaido, three in the main island of Honshu and two in the southern island, Kyushu. (Catchers attached to land stations usually move from one station to another, so do not all operate at the same time.) Japanese Government issues whaling licences to twelve catchers that are based at land stations.

Investments in Whaling

The EIU main report emphasises the declining position of the industry — "because whaling is not currently regarded as a developing industry, no significanct recent investments in whaling ships or other facilities have been made . . . figures show that two or three catcher boats have been built during the past three years, but no new mother ships have been built for many years . . . whaling companies tend to invest small amounts for renovating vessels in order to reduce labour requirements . . . " (see Future Prospects, page 58).

The Trade in Whale-based Products

Before the Japanese became involved in pelagic whaling, whales had been hunted almost exclusively for their oil. (The BWu was introduced as it set up a measure of how much oil could be obtained from each species of whale.) However, to the Japanese, whale meat for human consumption is the most important product from whales, though the revenue from her export of whale oil is not insignificant.

The total weight of whale products exported in 1972 was 33 890 tons, valued at 1 763 million yen. As a proportion of total exports, the value of whale product exports was approximately 0.02% in 1972. In 1971, it was roughly *0.08%* (EIU).

"The destinations of Japan's whale product exports in 1972 were as follows:

1.   **Whale Meat** — 2 173 tons were exported, mainly to the UK and the Ryuku Islands. This quantity was about 2% of Japan's production.

2.   **Baleen Whale Oil** — 29 586 tons of whale oil were exported to the Netherlands, the Republic of Korea, etc. This quantity was about 87% of Japan's production in that year. Whale oil is used as a raw material for producing margarine or shortening.

3.   **Sperm Oil** — this is now the main export. 2 059 tons of Sperm oil were exported to Holland, the Republic of Korea, etc. This quantity was about 7.5% of Japan's production. Sperm oil is mainly for softening leathers for gloves, soft clothes and shoe uppers, and for industrial use, particularly as an ingredient in cutting oils for machine tools. It probably provides just under 10% of the ingredients of these oils; a replacement oil could be used if sperm oil production were discontinued, but it is in regular use at present. Its other use is as an automatic transmission fluid in motor vehicles; however, this is on a small scale, and it is rapidly being phased out and replaced.

4.   **Canned Foods** — 47 tons of canned whale meat were exported, mainly to

Okinawa, the Federation of Malaysia, US Pacific Islands, etc."

Source: EIU report. This report, completed in June 1974, did not use the up-to-date data available on the trade in Japanese whale products.

Imports

The main whale product imported into Japan is whale meat. 25,477 tons at a value of 2,941 million yen were imported in 1973 (see Table 11). The main sources of imports were the USSR, Peru, Iceland, Spain and the Republic of South Africa. The pattern has changed a little according to import figures for the first 5 months of 1977, when 19,946 tons were imported at a value of $33.3 million. The USSR and Peru still headed the list of exporters, but next in significance came South Korea (a non-IWC member). Somalia (which was the registered home of the pirate ship Sierra), Iceland and Spain, with Brazil, North Korea and Norway at the bottom of the league. Soviet exports had risen to 85.4% of the total, Peruvian exports were both less in tons and as a percentage of the total; on the other hand, South Korea's exports had risen and three new exporters had appeared on the scene (Somalia, Brazil and Norway), of which Somalia, accounting for 2.5% of exports is not a member country of the IWC. Because the exports of whale meat have decreased, Japan has become a net importer, rather than a net exporter, of whale meat. But this merely means that her exports are of low grade meat and her imports are of high grade meat.

The Importance of Whale Meat outside Japan

In the immediate post-war years it was the policy of the British Government to encourage the consumption of whale meat and it was not subject to meat rationing. In the autumn of 1947, Lyons Cornerhouses were serving 600 whale steaks a day. But, as other meats became available again, whale meat became less popular. (The importation of whale meat is now forbidden.)

Outside Japan, whale meat is consumed in the USSR (no statistics are available) and in Norway, although much of the meat is given to the animals on their fur farms. And according to Dr. Jonsgard, a Norwegian whale biologist, "the whaling industry in Norway now hunts minke, killer and bottlenose whales on which the economy is most dependent for food for their minks". ("Banning the Importation of Whale Products in the European Economic Community and the United States", D. H. Petzel, Southampton College.)

Whale meat is also eaten in South America, largely in the form of 'fish' sausage.

Until very recently, Sperm Whale meat has not been widely used for human consumption, as the meat is very dark and contains a high proportion of haemoglobin in the blood and is unpalatable. However, methods have been found of reconstituting Sperm Whale meat. The Japanese, South Africans and the Russians now eat small quantities of this meat.

Table 11 – *IMPORTS OF WHALE MEAT TO JAPAN*

	1973			January-May 1977		
	Tons	*(%)*	*Value per ton (million yen)*	*Tons*	*(%)*	*Value per ton (US $)*
Angola	575	(2.3)	4.4			
Brazil				75	(0.3)	
Canada	535	(2.1)	3.2			
Central African Republic	50.9	(0.2)	0.2			
China	10.18	(0.04)	0.03			
Iceland	2,928.95	(11.5)	20.6	493	(2.1)	
Korea (North)	76.35	(0.3)	0.3	10	(0.04)	
Korea (South)	727.7	(2.9)	7.8	949	(4.0)	1,907
Norway				3	(0.01)	
Peru	1,577.9	(6.2)	2.8	972	(4.3)	730
Somalia				586	(2.5)	
South Africa	509.5	(2.0)	2.1			
Spain	1,959.85	(7.7)	8.8	292	(1.3)	
USSR	16,529.23	(64.9)	49.8	19.946	(85.4)	1,439

Source: Ministry of Finance, Japan (1973), Marine Mammal News (1977)

b)   FUTURE PROSPECTS

The picture that emerges from the above analysis is one of an industry that is in decline and is confused about its future. It is not surprising, therefore, that there has been little new investment in capital equipment in recent years and few developments in whaling technology. It is reasonable to assume that much of the currently-used equipment will soon be nearing the end of its useful life and will need replacing.

In this section we outline the kind of response that the industry might make to this situation.

1)   New Investment

Consistent with its own estimates of whale populations and its claim to be catching no more than a sustainable fraction of them, then the industry should make additional investment. The world food situation will ensure that in the short and medium term, at least, prices of meat will continue to rise and investment in equipment to improve catching efficiency would help to offset increasing costs. However, it is possible that the industry's estimates of the whales' prospects for survival when used to justify current actions, differ from those used to justify new investment.

It seems entirely probable that the industry will act for investment purposes on the estimates of the conservationist groups and will not invest significantly in new equipment. Certainly there have been no recent signs that the industry does intend to increase its investments.

2)   Retirement

The industry could, of course, allow a 'natural' death to whaling by phasing out equipment as it became too obsolete to operate profitably. Although perhaps a less attractive option from a strictly economic point of view, it could be chosen by default as both the Japanese and Russian industries wait for the other side to cease whaling first. This argument is quite often used to justify current whaling on the grounds that commercial viability will disappear long before whales become extinct; that whalers will become extinct before whales. It does not, however, take account of the fact that we do not know enough about whale population dynamics to judge at what point they will become extinct, or that the economic realities of such a situation, would encourage the annihilation of the species (see 4 below).

3)   Transferring the Assets

If it accepted the inevitable decline of whaling then the industry might decide to shift its assets immediately on the grounds that the earlier it opted out, the more value it could realise from them. This transfer could take two basic forms: the whole fleet could be sold to some other nation which might or might not use it for whaling, or it could be converted to other fishery uses or specialised roles. This option becomes increasingly attractive as the prospect of a total ban grows.

4)   Maximise Return

The fourth option open to the whalers is to slaughter as many whales as possible as quickly as possible within the lifetime of their existing equipment. A paper by Colin Clark on "Profit Maximisation and the Extinction of Animal Species" shows that on purely economic grounds and given the current high discount rate, that this is the best course to follow. This option has the added advantage of permitting asset transfers for non-whaling uses afterwards. In the latter case, the quicker the remaining whales are annihilated, the higher the return on the investment already made would be. An acceptance of the logic of this argument could lead to a variety of strategems by the whaling industry to avoid both existing controls and mounting pressures. They could, for instance, create front companies in non-IWC states and operate fleets under flags of convenience. Whale meat and other whale products could be processed so that they appeared on import figures under disguised headings, and so on. Previous experience in the trade in animal products in no way suggests that apparently reputable companies are above such practices when the economic logic encourages them. We are not suggesting that any of the existing companies are in fact engaged in such activities, but it would be naive to believe that they could not occur.[1]

The only option that guarantees some kind of future for whales is that of an immediate transfer of assets into non-whaling uses. All of the options outlined above are open to the Russians, plus an additional option. The Russian economy, being centrally planned, can make better long-term provision than a private company, in that it has a greater capacity to carry unprofitable concerns over lean periods. The Russians could therefore continue whaling, even if it became unprofitable, until such time as the Japanese companies were forced out of business by a combination of economic and political pressures. It would even be possible for the Russians to cease whaling temporarily in order to increase the political pressure on the Japanese. Were the Japanese to cease whaling, this would then leave the Russians effectively in control of the future of whales.

5) The Moratorium Option

It is relatively easy to understand why the whaling nations are opposed to a moratorium since it effectively puts an end to the industry. One can see that it might be very difficult to re-establish the industry after a 10 year hiatus, although the Russians might find this easier then the Japanese. If we take a world perspective, however, the moratorium option begins to look the most attractive future.

Dr. Sidney Holt has estimated that if whale stocks were given time to recover, whales could produce 2 million tons of protein annually — the world catch in the period 1962-64. The value from msy catches, with present utilization patterns and product prices, would therefore be about US $690 million, as compared with the 1972-73 actual catch value of US $170 million. If Sperm Whale meat could be made palatable this total could be 10-20% higher. Moreover, if catch quotas were established by weight instead of numbers (see page 72), although whales would have to be obtained from a relatively larger stock, the whaling effort would be considerably less.

Furthermore, the 10 year halt could provide time for more rigorous scientific studies of whale populations to be made and a realistic sustainable yield to be calculated. This, in turn, would allow a proper economic appraisal to be carried out and an appropriate level of investment to be determined. It would also provide the necessary opportunity for a total re-appraisal of the way whaling is conducted and controlled. Our feeling is that whales, if they are to be treated as a resource at all, are a common heritage resource that belong to all mankind and not just to those nations possessing the capacity to exploit it. We can see a case, therefore, for the future of whales being determined by all nations, not just those with a previous or current vested interest. It might be that the best future for whales and whaling would lie in it being taken completely out of the hands of sectional private and national interests. We would advocate the establishment of

---

(1) These arguments were confirmed in an article in the Observer Magazine, 24th November 1974.

an international agency, funded initially by all countries in proportion to their previous benefit from the use of the resource, which would not only control whaling but would also purchase and operate the whaling fleets.

If one of the major arguments for whaling is to be the contribution that whale meat could make to alleviate protein deficiency, we would suggest that such an agency would not be bound to operate on a purely economic basis, but would be charged with assisting first those countries suffering from protein deficiencies. Such a scheme would, of course, establish a great many precedents for the control and exploitation of all common heritage resources. We cannot see that any of those precedents are to be feared except by the greedy and the short-sighted. We do not view with optimism the likelihood of such an approach being adopted. Nothing in the history of man's exploitation of whales suggests that the rational will prevail.

CHAPTER 7

ANALYSIS OF THE SCIENTIFIC BASIS OF WHALING

Some people have argued that whaling can and should continue, as the scientists now know exactly how many whales can safely be taken without harming the populations. We suggest that this is not true. We are alarmed by how little the scientists do know about Whales, especially social structures and migratory patterns and by the unreliability of the methods used to assess population sizes. We do not wish to underestimate the work of many very competent whaling scientists, but we feel that there is absolutely no justification for their complacency. Our analysis of their often conflicting findings leads us to the opinion that insufficient reliable data exists to make confident predictions about the number of whales that can safely be killed.

There is a growing feeling that many scientists are failing to consider the social and political uses to which their findings can be put. Whaling scientists have not been exceptional in this respect and there have been occasions on which the commercial interests have been able to turn incautious statements to great political advantage. What debate and criticism of conflicting opinions there is goes on behind closed doors or within the covers of scientific journals that are, for the most part, unread by, or inaccessible to, the public.

In the arguments we have put forward we have not sought to produce any single alternative judgement. We have merely attempted to demonstrate the wide range of opinion that exists and the likely errors involved.

a) DATA COLLECTION

One of the most striking features of whale research is the enormous data bank that has been accumulated by the Bureau of International Whaling Statistics (BIWS) and others, and the lack of use that has been made of it. There are many analyses of this data that could be made which would extend our knowledge of whale populations. However, there are few scientists in this field and, as they have limited resources, much of the data remains to be interpreted. Even if there were enough scientists, much of the data which they would need is not published and is only available to the industry. This is particularly true of catch per unit effort data (see page 63).

In 1974 a FAO/ACMRR Working Party on Marine Mammals produced a paper which examined the problems of cetacean research and many of the points they raised are itemised below:

1) Unpublished Data

Many of the calculations on which the Scientific Committee of the IWC based its advice remain unpublished and, with the passage of time, it is now difficult to establish how they were derived. It is, therefore, impossible for the layman or scientist to examine the calculations objectively without knowing the methods and biases involved.

2) Availability of the Data

Not all the data is available and not all the whaling countries make their complete data available to the BIWC or to the IWC. This is particularly true of the USSR, who do not make their complete biological or marking data available to the Commission. This means that almost half the marking data is missing.

3) Methods of Estimating Population Sizes

Populations of whales are usually estimated by a combination of methods:
    a) Sightings, which literally means seeing whales.
    b) Catch per unit effort (CPUE)
    c) Marking or tagging.

a) Sightings
There are three main reasons why population estimates based on sightings are suspect:
i)   Because the sightings are usually made by the whaling industry in the course of their operations. Whilst not wishing to imply that there is any deliberate misinterpretation of the facts, one can see obvious commercial advantages to the whaling industry in concealing the true state of affairs.
ii)   Because the whalers go to the places where they expect to find whales. Whales are not distributed evenly throughout the oceans, and although certain allowances are taken into consideration when extrapolations on global population sizes are made, these allowances can only be suspect as it is not known how many whales there are elsewhere.[1]
iii)   The chain of calculation of numbers from sightings is so complex that errors at each stage are compounded such that a reliable estimate is extremely unlikely.

b) Catch per Unit Effort (CPUE)
CPUE means counting the number of whales caught and relating the count to the amount of effort it takes to catch them. For example, if it takes a given fleet a particular 'effort' to harvest a certain number of species in a given period of time, it is considered that the population is stable. The total world population of the species is then calculated on this basis. If it takes more time and effort to catch the same number of whales, then it is assumed that the population is declining. What is not always taken into account is

---

(1) "While elaborate calculations are used in an attempt to reflect different densities of whales in different parts of the ocean, the fact remains that the observations are not random samples — the whaling fleets are where the whales are — and any estimates based on such biased sampling technique must be extremely suspect." Interview with Dr. Lee Talbot, Oct. 11 and 15, 1973.

the increasing efficiency of whale-killing techniques (e.g. the use of ASDIC — a type of sonar), so that the 'effort' rarely stays stable. Thus it is possible that two identical whaling fleets applying the same effort could produce different catches and therefore any calculations based on these figures could give two different population estimates for the same species.

Dr. Sidney Holt's FAO/ACMRR Working Party Report suggests that the greatest danger in the CPUE system is this bias from inappropriate measures of effort, and that errors come from:

"i)   unaccounted for changes in catcher efficiency, as individual vessels and as fleets,

ii)   the effects of cooperation between expeditions, and division of labour among catcher/scout vessels,

iii)   differences in performance by catchers from different national fleets; the relative efforts exerted which change through the years, reinforced by the incompleteness of the data,

iv)   changes in the area of distribution of the fleets of different countries,

v)   changing 'interest' in different species of whales,

vi)   responses to restrictive quotas and to national or expedition allocations of quotas,

vii)   systems of payment of gunners and crews, differences in which change motivation with respect to species and sizes preferred, and

viii) changes in the fraction of a catcher's operating time spent searching or chasing, varying, among other things with the number of whales, of all species taken, and

ix)   the contraction of the 'area occupied' by a whale population as the species is depleted."

c)   **Tagging or Marking**

Many animal populations have been studied using tags and marks, the most widely applied being bird-ringing (banding). The use of numbered tags is one of the safest and surest ways of obtaining the data necessary to construct accurate models of animal populations and their changes. The type of whale mark used at present is in the form of a metal cylinder fired from a gun into the blubber of the whale. It is recovered either at the flensing stage or from the oil boilers. Unfortunately there are a number of serious difficulties involved with the use of tags on whales. These include:

i)   Cost — in order to implant the tag, a boat and expert crew is needed. We believe that this difficulty is often exaggerated. Intensive programmes with the cooperation of other research interests as well as the industry could overcome much of the expense problem.

ii)   It is difficult to be certain that the mark is successfully planted. Recent work with visible streamers attached to the tags will help overcome this problem to a large extent.

iii)   Recovery — the efficiency with which the tags are recovered is variable. It is often not possible to establish which whale a tag has come from when the tags are only recovered from the blubber boilers. This considerably reduces the usefulness of the data. The differences between the efficiency of recovery of the various fleets are also quite marked.

iv) Lack of information – this is a fault for which there is no excuse. The USSR does not make available its base-line data on how many tags have been planted in which species; only data on the recovered tags are published. This means that it is impossible to attempt calculation of mortality rates.

v) Injury – used properly the tags do not harm the whale, but it is possible that injury is caused occasionally; there is no information, however, on the mortality due to tagging.

vi) Numbers tagged – this is related to cost, but it should be stressed that the numbers marked are so small in proportion to the catch as to make any assessments based on the marking programme only very sketchy indeed.

The following points are important in data collection, which we believe are not adequately taken into account:

4) Density

The social structure of the whale populations is hardly understood, but there is evidence that some species of whales (and other animals) tend to maintain their population density by contracting the area (or territory) occupied. This will affect the sightings and catch rate in exploited areas, obscuring the decline in numbers.

5) Behaviour

The behaviour of whales has only been superficially investigated and needs to be taken into account when estimating populations. For instance, Minke Whales often approach catcher boats, while other species disperse.

6) Infractions

It is unlikely that whalers always declare infractions of IWC rulings and this must introduce a distortion of the data. That whalers do commit infractions is proved by the fact that they have returned markers planted in protected species. It is also known that undersized whales are often 'stretched', i.e. that animals just below the size limit are given the 'benefit' of the doubt by the whalers. The Commission is complacent in its assumption that all infractions are reported by the whalers to the Bureau of International Whaling Statistics.

b)   ANALYSIS OF THE DATA

Population Dynamics

The structure of, and changes in, whale populations are very sketchily known, but the theoretical possibilities as derived from studies of other animals are outlined below.

The size of a population is governed by the surviving young (recruitment) and the deaths (mortality). Factors that control recruitment are: the fertility of the mothers, survival to maturity, the age at which the females mature, and the percentage of females in the population. Factors controlling mortality include the food supply, the competition for space and food, the numbers of predators, parasites and diseases, and climatic factors.

Controlling factors are classed as 'density dependent' and 'density independent' according to whether or not the numbers vary directly with the density of the population. Density independent factors are, for example, bad weather, the extent of suitable habitat and in some cases the amount of food available. Density dependent factors act as a form of internal check. They can include behavioural and physiological mechanisms, such as defence of territory, forcing out weaker individuals into less suitable areas, decreased fertility in females in crowded conditions and increased fertility when numbers are low.

Population of mammals generally do not increase to the limit of their food supply, and so food becomes a limiting factor only when a population has expanded dramatically. However, it is anticipated that the present experimental harvesting of the food supply (krill) of Antarctic baleen whales will become a major industry in the forseeable future, thus preventing depleted stocks from recovering and threatening others. Meanwhile, estimates of whale populations correlate so well with man's activity that killing is unquestionably the chief factor affecting their numbers.

Natural predators are not generally thought to do more than control some of the surplus in the population, as their numbers appear to follow the changes in numbers of their prey. Predators may reduce the violence of the change, by preventing populations of prey from exploding at a compound interest rate (more breeding females — more young — more breeding females); they are not normally capable of exterminating the prey in a naturally balanced ecosystem. The case of man and whales breaks the rules, as man has demonstrated by hunting several species to commercial extinction.

Types of Populations

All animal populations vary in numbers with time; some vary within a narrow range, others show large fluctuations. In an undisturbed, established system the average numbers for any one population will be the same as long as the average is taken over a long enough period of time. If the numbers do not vary greatly and the changes are quick, a population might be described as stable; if the numbers show large changes and only correct themselves over time intervals of several generations, the population can be called 'fluctuating' — typical are the lemmings, voles and Peruvian anchovy. In these animals large crashes in numbers occur. The cause of crashes in fluctuating populations are not always known, though several theories exist to explain them; including exhaustion of food or of an essential nutrient, or selection for more aggressive, less fertile individuals as density increases.

The above theories on population dynamics and structure were based on data drawn either from laboratory studies or on observations in the field on animals in some kind of ecological balance, not predated by man. Whale populations as they exist at present cannot be regarded as typical of animals in a natural balance with their environment, subject as they are to heavy predation by man.

Structure of Populations

A stable population, not heavily exploited, would show a fairly consistent structure: the proportion of the different ages will tend to remain constant.

However, knowledge of whale population structure is based on data gathered since the start of intensive whaling and therefore mainly reflects the structures of populations under a changing pressure. Density-dependent factors in a heavily exploited fish population can encourage increased fertility, with a consequent production of more young animals. It is uncertain whether the example can be extended to marine mammals; the extra number of young whales that can be produced is very limited, compared for instance with fish, where egg production is enormous. Of young whales born, it is possible that not more than 50% survive. The whale has been under pressure for some time, and the present structure, showing an abundance of young animals, reflects the pressure on the population. However, Dr. P. E. Purves has suggested that the apparent age structure may reflect a population which is slowly increasing with very low adult mortality. Dr. R. Laws concludes that, although there appears to be an increase in the **proportion** of young, the signs are that the populations are decreasing.

Culling a Population

The arguments have been put forward that:
1)   A 'maximum sustainable yield' of whales exists, and is a desirable goal for whalers.
2)   It will be beneficial to cull certain whale species because they are taking food from the mouths of endangered whale species (endangered by over-fishing).[1]

---

(1) "Harvesting of whales was still necessary", Dr. Gambell added, "because the unchecked growth of the whale population would mean food competition between the species. For instance the Minke (which was introduced to the quota system three years ago) would eat the zooplankton (krill) which resembles a 2½ inch shrimp. This tiny animal is the basic food supply of the Giant Blue, weighing 100/150 tons and larger than 30 elephants." Asahi Evening News, 24.6.74

The first argument is based on a neat, abstract concept which no self respecting ecologist would dare to apply to a species as little understood as the whale. The second denies a widely accepted hypothesis — that two species cannot be identical in all their requirements and survive together.

Taking them one at a time:

1)   The maximum sustainable yield (msy) theory is as follows —

Any one breeding stock of whales has a stable population level, and tends to return always towards that level. If you kill a number of whales from that stock, the remainder will put their energies into returning to the original number — they will produce more young, by maturing sooner and becoming pregnant more often. (They can only do so, incidentally, if there is sufficient extra food for them.) If you continue to remove precisely the right number of whales, you can hold the whale's population at a steady level, and because it is trying to get back to its original population level, you will always have some excess whales to cull. At a certain level of population — called the msy stock level — the number of whales you can take from the population without causing it to decline is at its highest possible for that breeding stock of whales.

Thus, for different whale species, the scientists have developed mathematical models of what they hope is the relationship between the numbers of whales in the stock and the excess they may safely allow to be taken. In practice they say — if the initial population of this stock of Sei Whales was X thousand, we believe that you get an msy when the stock is at Y thousand, and that msy will be Z thousand or hundred. They normally take msy stock level (Y) to be approximately 60% of the initial population level. As shown elsewhere, they had great trouble estimating what the present numbers of whales are, let alone what the initial numbers were; secondly, the mathematical relationship between initial numbers and so-called msy level is a matter of assumption rather than fact; and third, the theory relies on density dependent factors operating within the individual breeding stocks of whale species. There are two hitches to the third assumption: one is that information as to the identity of separate breeding populations is extremely poor, so that whales that are managed as one population (given a single msy) may well consist of several different breeding groups. Even if the mathematics reflect reality accurately, if applied to several stocks when they are believed to be one, considerable error will occur. The second hitch is that it is questionable in the case of Sperm Whales whether any density dependent factors operate.

A Working Group of the FAO/UNEP Scientific Consultation on Marine Mammals (1976) reported that "it was thought to be both difficult and risky to use this (msy) as a management goal because the level giving msy is difficult to estimate, and because this level fluctuates due to environmental variations, both physical and biological, and because it ignores the complex of social costs and benefits relating to harvesting. Reductions of the populations to such levels are risky, as discussed above, and especially in practice when abundances are poorly estimated."

A paper to the Scientific Committee of the IWC in 1977 (SC/29/Doc.6) has taken this idea further and shows that as a population approaches msy level (assuming that to be accurately known, that the stock managed is a separate breeding stock and that calculations of numbers of whales are precise), it becomes increasingly vulnerable to fluctuations resulting from environmental variations, both physical and biological. Stability of the population, as measured by its return time, drops almost exponentially. The consequence of the ideas raised in this paper is that if a whale population at or near msy level were hit by some external factor, recovery might be impossible, or might take a very long time, far longer than if the population were at, say, 80% of its initial level.* The paper suggested that to allow for this risk, quotas for baleen whales in the sustained management category (see p.33) should be set at less than the current 90% of msy. In 1977 a majority of the Committee felt that insufficient information was available concerning the effects involved to warrant an immediate change in the level, relative to msy, used in determining catch limits.

This seems a strange conclusion overall, since the above analysis assumed among other things that the msy level was accurately known. Yet a glance at the Scientific Committee's own estimates of msy reveals amazing variations with time. In the case of Southern Hemisphere Sei Whales the difference between the msy estimates made between 1967 and 1976 for the same group of whales is larger than the actual quota set. The quota is set at 90% of msy — clearly when the msy estimates vary by more than the actual quota a 10% safety margin is not enough. Taking this, and the previous considerations into account, the only safe safety factor is likely to be 100%.

Let us also look a little more closely at the theory. It is concerned purely with a numerical relationship between original numbers of whales, increased fertility and recruitment to the stock at lower numbers, etc. However, whales have a complex social structure which may be altered or destroyed by heavy fishing pressure. It is now accepted among ecologists and behaviourists that social factors affect the emotional well-being of a population, which in turn affects its productivity and therefore its survival. If, as the model builder claims, fishing increases the numbers of young, the social structure will be affected, probably causing a decrease in the final numbers of surviving young. Recently the IWC has altered its model for sperm whale msy by conceding that the so-called non-breeding males may have a role to play in the survival of the

---

*In 1965 Chittleborough estimated that it would take nearly 50 years for the Area IV Humpback Whale to recover to even 50% of its initial unexploited size. "Right whales have been protected for long periods in all major oceans but as yet there is only a slight encouraging sign that the Southern Ocean and North Atlantic populations are beginning to recolonize their old ranges; of the recovery of the North Pacific stocks there is no sign at all. Recent observations in Areas V and VI (Machinda 1974) give no cause for optimism regarding recovery of Blue Whales or Humpbacks in those sectors . . . are our present ideas on the rates of baleen whale population recovery from very low levels just so much wishful thinking". (Gaskin, Oceanogr. Mar. Biol. Ann Rev. 1976, 14, 247-346.)

species, and reducing thereby the number of whales that the whalers could kill. More factors have almost certainly been left out, simply because they are not yet measurable.

Perhaps most seriously, the model assumes that the whales have naturally stable populations. If this is not so, and it is highly improbable, then the biomass of whales removed will need to be varied with the fluctuations. The Peruvian anchoveta fishery may well be a case in point. Periodically this fish goes through a population crash. Fishing effort, however, intensified so that on the occasion of a natural crash, an added stress affected the fishery and the following year, when the population would normally have been recovering, it was set back by further fishing effort. It is not yet known if it will recover and both fish and local fishing fleets have suffered as a result of lack of restraint. Now our estimates of whale populations are largely based on catches and we have no way of knowing if a natural crash has occurred until the boats have come back again. We would challenge the model builders to produce a model of msy consistent with fluctuating populations (see page 108).

Finally, despite estimates of msy on weight for sperm whales, a logical approach for an industry, the Commission has rejected it because it meant lower quotas. This shows clearly that the industry is only interested in this year's quotas, not in a sustainable yield of products.

In summary, there are many areas wide open to error in calculating the maximum sustainable yield: estimates of populations, type of function relationship chosen, assessment of long term environmental or social effects. The concept itself may well be invalid — in an ecological sense it most certainly is since it takes no account of interactions between linked species in the system. The IUCN in its 1975 General Assembly passed a resolution to the effect that msy was 'not adequate as a basis for management of wild living resources' and that management should instead be based on ecological relationships according to principles stated elsewhere in the resolution.

The msy concept has been developed purely as a response to the problems of restraining governments from overexploiting. It is not therefore surprising to find that it has not been employed by population theorists, nor is it surprising that to our knowledge no-one has claimed to be able to calculate a workable msy for better known species than whales. The New Management Procedure of the IWC claims to protect whale stocks that are in need of protection — yet there is no estimate of the critical survival levels of the various stocks of whales. Below that level the population will not survive, simply because the reproductive rate is so low that the population cannot increase sufficiently rapidly and one set-back is enough to wipe it out. Without an estimate of that level, there is no way by which the IWC can protect whale stocks 'when they need protection'.

2)   The second point made is that because some whales may compete for the same food (krill), it may be beneficial to kill more of the more numerous species

to allow the over-exploited ones to recover. This argument has been used to support the killing of more Minke whales because they are said to compete with the Blue (which Britain caused to become endangered), and is now being used to encourage killing of other baleen whales which compete with the Blue and the Fin. We should first point out that on theoretical grounds this argument is shaky. All species of whales have evolved together over thousands of years in ecological balance, so that a high degree of competition for food is not likely (it is a generally accepted working theory that no two species can be identical in their requirements for space and food). Exploited populations may find a new and possibly different balance with more numerous species — but there is no way that we can say that exploiting, for instance, the Minke will help the Blue.

Indeed, it is likely that it will only endanger the Minke. Any scientist who chooses to use this argument to support higher quotas for Minke whales is guilty at best of stupidity, at worst of dishonesty. It is by now well known that Antarctic seals are increasing at a formidable rate — and that they consume more krill than the Antarctic baleen whales put together. Since they have a faster generation time than whales, it is likely that seals, rather than whales, may be causing problems for the recovery of protected whale stocks.

Finally, it is now impossible for any scientist to claim that whales and seals are the only competitors for krill. In addition to the penguins and seabirds who eat krill to a greater or lesser degree, man is now entering the field as a serious competitor. Some half-dozen countries have declared their interest, and a number — Japan, Russia and West Germany among them — have sent out exploratory fleets (see pages 137-140). If mankind starts to exploit krill on a large scale, most of the whales, many of the seals and other Antarctic species are likely to suffer. No amount of killing Minke will save the Blue Whale then.

c)   CRITICISMS OF OTHER WAYS IN WHICH DATA IS PROCESSED

1.   'Best' Estimates

The Scientific Committee, in the early days, when faced with its members having different opinions, presented the Commissioners with a range of estimates. The Commissioners naturally acted — if they acted at all — on the **highest** ones. To counter this the Scientific Committee decided to arrive at all costs on a single estimate ("the best we can do now, with all the available data, methods and time"), recognising there was a degree of uncertainty in it, and they might, with new data and methods, have to change it next year. The Scientific Committee also tried to set limits to their uncertainty, "if the best estimate is x, and it is almost certainly not greater than y". This was a great improvement, but it rebounded. When little was known about a newly exploited species — Sei or Minke — they would agree, temporarily on only very vague statements; in other cases, while giving a "best" they had to admit that some other very different one — much higher or much lower — could almost equally well be consistent with the data. And the methods available do not yet permit calculation of the statistically **probable** errors, which would have made possible a semblance of scientific decision-making. In these circumstances, we do not see how any safe estimate can be made.

2.   Age of Sexual Maturity

Much emphasis has been placed on the apparent reduction in the age of sexual maturity of female whales (based on the pattern of layers in the earplug) and on the consequent assumption that whale populations will therefore increase at a faster rate. Theoretically, when a whale population increases because of changing factors such as the availability of food (possibly induced by killing off a proportion of that population or by the exploitation of another species), the young whales grow larger more quickly. The whales become sexually mature at a certain size rather than age and, as they get bigger younger, they become sexually mature younger. Yet there is still some doubt about the significance of such changes and how universal they are; the real importance of these changes is still open to interpretation. Many questions still have to be answered. For example, because the whales reach sexual maturity earlier, it does not necessarily follow that they breed earlier or are capable of looking after their offspring. The highest infant mortality rates might occur with the young inexperienced parents, whilst the large proficient breeders are the preferred targets of the whaling industry.

3.   Biomass (total living mass)

The FAO/ACMRR Working Party drew particular attention to the importance of assessing the weight as well as the numbers of whale populations. For instance, under the BWu system, which was based on the amount of oil a whale yielded, by catching Sei Whales (1 Bwu = 6 Sei), the whalers could obtain 50% more whales by weight — for meat. Purely numerical assessments of populations do not take into account the decreases in overall biomass of the population, and biomass is ecologically more significant than numbers. Furthermore, the whaling industry has naturally always measured its success by weight and price of its products, not by the number of whales caught. The decline in overall biomass has been noted in nearly all exploited whale stocks. The total biomass of baleen whales in the Antarctic and of Sperm Whales in the world has been reduced to levels far below one third or even one quarter of their initial levels.

The concept of sustainable yield in weight is not new. It was introduced in the early 1960's — at the same time as the criterion of msy in numbers. (NB in 1960 an Ad Hoc Scientific Sub-Committee reported to the IWC meeting that there was a "progressive reduction in the sizes of Sperm Whales from before the war up to the present time" and that they regarded this "as a warning that the Antarctic stock was being appreciably affected by whaling".) If one accepts the methods used for calculating msy, then the msy of a species by weight would be obtained from a relatively larger stock than would the msy by numbers, and with a lighter rate of exploitation. As the FAO report explained, "setting the ultimate objective of maximising the yields in weight does not imply that regulation need be by quotas of weight rather than of numbers; eventually numerical regulations and size limits could be adjusted to ensure an appropriate level of whaling effort to maximise the product weight or value. This objective is also more 'conservative' in that emphasis would be placed on catching the older and larger animals, at a lower overall rate of exploitation and with less risk of inadvertent depletion of the stock to levels at which reversibility or even its survival, are threatened."

Because the Scientific Committee to the IWC has only been concerned with the interests of the Commission, little work on biomass has been done even though the basic data are available.

4.   Irreversibility

When two or more species compete for a limiting environmental factor (such as food), and one species declines, the decline can often be irreversible. Several whale species are thought to be interdependent at least to some degree and so the depletion of one may in some cases lead to an increase of another. For example, Blue, Fin and Sei stocks are thought to be inter-dependent in the southern hemisphere. But the population recovery (growth) rates — if they occur — are so slow that it would seem very unwise to attempt any predictions of population recovery rates.

5.   Time Lag

As the FAO/ACMRR paper pointed out, "The time lags between intensifying exploitation, scientific evidence of the effects of this, and imposition of minimum regulations are such that practically without exception the stock is overfished before the exploitation is adequately controlled. This was true for all the large baleen whale species, and also for Sperm Whales."

CHAPTER 8

THE CRUELTY ASPECTS OF WHALING

The cruelty involved in killing whales is thought to be considerable, even under ideal conditions. The weapon used to kill whales is the explosive harpoon, which was originally invented in Britain in 1840, developed to its present level in Norway by Svend Foyn and over the years has seen considerable changes. Modern harpoons weigh about 160 lbs. and are 6 feet long. They have four-barbed warheads fitted with a grenade which is exploded by a time fuse inside the whale. They are propelled from a gun and reach over 60 mph on impact.

Diagram 2

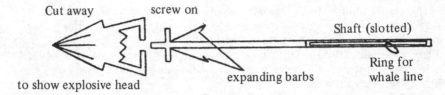

Cut away
screw on
Shaft (slotted)
to show explosive head
expanding barbs
Ring for
whale line

The process of harpooning whales was described by Sir Alister Hardy, the eminent oceanographer, as, "if herds of great land mammals, say elephant or buffalo, were chased in armoured vehicles firing explosive grenades from a cannon, and then hauled close at the end of a line and bombarded until dead".

In order to cause instantaneous death, the harpooner must fire the harpoon at a point just behind the head, but in the rough conditions of the Antarctic and the open seas this can only rarely be achieved. More often the whale is hit in the abdomen or other parts of the body, causing a prolonged death, and possibly requiring additional harpoons. In a fact sheet published by the Japan Whaling Association in June 1974, it was claimed that an average of 1.5 harpoons were needed to kill a whale. This means that a large number of whales do not die anything like instantly; when struck by the first harpoon they will sound and then surface at some distance from the boat. The whalers, and some of the scientists who support them, claim that nowadays whales are usually killed quickly and humanely, but one wonders if they would consider three minutes (the time claimed as an average minimum) as a humane time if the same standard were applied to themselves; we certainly do not consider 3 minutes humane when slaughtering other livestock.

It has been well established that whales have a highly developed nervous system, in many ways comparable to land mammals and therefore almost certainly feel pain in a comparable way. Applied to man or even his domestic animals, such a death would be considered absolutely outrageous in most societies (Japan being a possible exception as it was only as recently as April 1974 that animal protection laws were introduced).

"If we could imagine a horse having two or three explosive spears driven into it, and then made to drag a heavy butcher's truck while blood poured over the roadway until the animal collapsed an hour or more later, we should have some idea of what a whale goes through." Dr. Harry Lille, ex-ship's surgeon.

ALTERNATIVES TO THE EXPLOSIVE HARPOON

At its 27th meeting in 1975 the IWC accepted a recommendation of the Scientific Committee that it should make enquiries about possible new developments in chemical and explosives suitable for killing whales; examine ways of improving the efficiency of existing methods, including the killing of small whales where explosives cannot profitably be used; and look into the training of gunners.

In addition, the Commissioners for whaling countries, and countries until recently engaged in whaling, were asked to provide details of any developments or research undertaken in the field of humane killing since the Commission's Working party reported in 1959 and of any investigational work now in progress.

Research has been carried out in America on the drug etorphine hydrochloride and its possible application to the whaling industry. This drug is marketed in the UK under the trade name Immobilon by Reckett and Colman. Its effectiveness for whales is questionable as it takes 2 minutes or more to take effect, during which time the animal could dive and might sink through being unable to retain air in its lungs.

There are two further objections to the use of the drug; first, its very potency makes it extremely hazardous to use, especially under field conditions in the whaling industry, and second, the drug would normally remain as a residue in the whale's flesh for at least one week; it is unlikely therefore that health authorities would allow the import or sale of whale products for human consumption from animals killed by etorphine hydrochloride.

Tentative and inconclusive studies have also been done on the use of a chemical tranquilizer MS222 SANDOZ, at present in use as a fish anaesthetic. As no research has been carried out on whales or on land animals of a comparable size, it is exceedingly difficult to feel at all sanguine about the future of this drug in whaling.

Valuable research has been done on the delivery by projectile of biologic restraining materials to domestic animals.* Unfortunately the range of animals and situations tested was very limited and without a great deal more work it is difficult to imagine that it will be relevant to whaling.

Several years ago a British company, Hector Whaling Co., spent over £280,000

---

*"Ballistic Delivery of Biological Reagents" Howard A. Baldwin and Charles S. Williams. Final Report prepared for Emergency Programs, Animal and Plant Health Inspection Service, US Dept. of Agriculture, September 1973.

on experiments lasting some years in an attempt to find an effective and humane method of killing by electrocution, but these experiments were not successful. Our view is that since then not nearly enough money, time or research has been employed in finding more humane methods of killing whales. At the 28th IWC meeting in 1976 the subject of humane killing was again raised. The Secretary has been asked to enquire into what research is currently being undertaken and to ask again about better ways of killing Minke and other small whales.

CHAPTER 9

ACTION GUIDE

a) BACKGROUND

The previous sections of the manual have detailed a record of the shortsighted and insensitive relationship between man and whales. The British share of this insensitivity has been considerable. We helped to kill off the Blue Whales of the Antarctic and North Atlantic, and the North Atlantic Right Whale, as well as having a hand in developing hunting technology such as the powered harpoon and fast boats for catching.

To our credit, Britain did at least make a worthwhile gesture in attempting to control catches of the North Atlantic stocks of Blue Whales taken by *the United Kingdom*, Germany and Norway before the Second World War. Similarly, Britain pulled out of the Antarctic quite early, but not before having decimated the Fin and Blue Whale stocks. However, the decline of British whaling did not signify a decline of interest in whale products which she was able to utilise to great advantage.

In the 1960's the British Government's attitude at the IWC meetings began to change; whereas in the past it had objected to lower quotas, the British now began to support more rigorous conservation measures strongly, thus mitigating, to a limited extent, their earlier errors. Whilst recognising that the British must accept blame for past actions, this in no way justifies present Russian and Japanese actions.

Britain has done relatively little to redress the damage of the past. Although the gesture of voting for the 10 year moratorium at Stockholm was made in June 1972, it was not until March 1973 that the partial import ban on baleen whale imports was introduced. And even though Britain has voted for the moratorium on each subsequent occasion (1972 and 1973 IWC meetings), there are many possible actions that the Government has failed to take.

In spite of the Government's lack of action, many MPs of all political shades were actively supporting the whales' cause. Sir Charles Taylor held an adjournment debate on whales on 1st May 1972. A few months later Sir John Langford-Holt introduced a Private Member's Bill to ban the import of whale products, but this was voted down in June 1972. The Earl of Selkirk also *initiated* a debate on whales in the House of Lords on 2nd August 1972. During the summer of 1972, six MPs — David James, Peter Hardy, Sir John Langford-Holt, Peter Archer, Ernie Money and David Clark — sponsored Early Day Motion 404: "That this House, noting that the International Whaling Commission has failed to implement the Stockholm Conference decision to put a 10 year moratorium on whaling, once again urges Her Majesty's Government to prohibit the import of all whalemeat and whale products". This motion was signed by over 200 MPs.

Petition to:
Prime Minister Tanaka and Premier Kosygin

The history of whaling is a sad one. Before the International Whaling Commission was set up in 1946, the UK, USA, Holland, Spain, Norway and many other countries were killing thousands of whales each year without any quotas or controls. They apparently did not consider that their actions could lead to the extinction of whole populations of whales. Predictably, it did: the North Atlantic Right whale and the Atlantic Grey whale disappeared altogether, and the first thing the IWC did when it was established was to give total protection to the other species of Grey and Right whales whose stocks had become severely depleted by over-exploitation. However, the hunting of Humpback and Blue whales was still permitted despite precariously low numbers.

We realise that neither Russia nor Japan were responsible for the destruction of the Grey and Right whale stocks as both countries have only become involved in whaling on a large scale comparatively recently.

Having a high regard for the culture of the Soviet and Japanese people, we are convinced that they do not wish to be responsible for the extinction of any whale populations. Catch controls have consistently been based on the most optimistic figures when the only responsible procedure would have been to err on the side of caution. Unless the 10 year moratorium proposed by 122 nations at the Stockholm Conference is introduced to allow the severely depleted whale populations to regenerate, we fear that the Fin, Sei and Bryde's whales will also become commercially extinct.

Although Sperm whales are still relatively numerous, we are concerned for their future as the other species become less commercially exploitable.

Whale products are of considerable economic importance but they are not vital; nor is whale meat indispensable to the diet of the people of Japan or Russia. Further, we think it questionable whether Sperm whales should be killed for the sake of the Soviet space exploration programme where we understand sperm oil is used as a high temperature lubricant. There will be little credit to man if in the process of getting to Mars he also wipes out the whales. We understand that adequate substitutes exist for all whale products.

Certain aspects of whaling can only be described as barbaric. We think that in the name of human dignity the killing should stop, at least until a much more humane capture technique has been devised.

We believe that the fate of the whales should be determined by all nations and not just by those countries with the advanced technology to kill them in bulk.

Therefore, we the undersigned, urge Prime Minister Tanaka and Premier Kosygin to bear in mind the high importance of public goodwill in matters of international trade, and the danger to other industries of unfavourable world opinion: to take heed of the wishes of the overwhelming majority of nations who expressed their views quite positively at the 1972 Stockholm Conference on the Human Environment: and to take immediate steps to phase out whaling.

Peter Balfour, Chairman & Chief Executive, Scottish & Newcastle Breweries Ltd.

R. M. Bateman, Chairman, Turner & Newall Ltd.

Sir Raymond Brookes, Chairman & Chief Executive, Guest, Keen & Nettlefolds Ltd.

I. O. Chance, Chairman, Christies, Manson & Woods.

Frank Chapple, General Secretary, Electrical, Electronic Telecommunication & Plumbing Union.

R. H. Gane, Chairman, George Wimpey & Company Ltd.

James M. Goldsmith, Chairman, Cavenham Ltd.

Charles Hambro, Chairman, Hambros Bank Ltd.

Clive Jenkins, General Secretary, Association of Scientific, Technical and Managerial Staffs.

Oliver Jessel, Chairman, Jessel Securities Ltd.

John O. Lyle, Chairman, Tate & Lyle Ltd.

Rt. Revd. Hugh Montefiore, Bishop of Kingston.

Sir Arthur Norman, Trustee, World Wildlife Fund, British National Appeal.

T. J. Palmer.

D. E. A. Pettit, Chairman, National Freight Corporation.

J. S. Rejeange.

Le Baron Guy de Rothschild, Chairman, Banque Rothschild, Paris.

Alan Sapper, General Secretary, Association of Cinematograph, Television and Allied Technicians.

Hugh Scanlon, President, Amalgamated Union of Engineering Workers.

L. E. Smith, Chairman, The British Oxygen Company Ltd.

G. G. Stockwell.

Lord Thomson of Fleet, Chairman, The Thomson Organisation Ltd.

Le Baron Gérard de Waldner.

H. A. Walker, Chairman, Bass Charrington.

79

One is killed every 20 minutes. Is this carnage really necessary?

An open letter to the International Whaling Commission meeting in London today.

The International Whaling Commission was set up by 14 whaling nations in 1946 to safeguard for future generations "the great natural resources represented by the whale stocks" and "to protect all species of whales from further over fishing".

We believe a 10 YEAR MORATORIUM ON ALL COMMERCIAL WHALING should be introduced at the IWC meeting, as recommended by the United Nations Conference on the Human Environment held in Stockholm last June, for the following reasons:-

1. To allow severely depleted whale populations to regenerate. Five species of whales (blue, humpback, bowhead, right and gray) are now so reduced in numbers commercial exploitation is not profitable. As a result these species have been given total protection. The Fin Whale (second biggest in the world) is at a low ebb, but still being exploited. Protected species are still caught by non-members of the IWC in the course of hunting more numerous species.

2. The only justification we can see for killing whales would be to use them as a protein source for hungry people. Few countries still import whalemeat for human consumption, and even in these instances the use is declining for social reasons. Whales are not indispensable to the human diet, nor are they essential to industry. Present uses for whale products – petfood, margarine, cosmetics and transmission oil – are all covered by acceptable substitutes.

3. Whales exist in international waters; no one country (or even two) should assume, as they have done, the right to over-exploit them. In the past the whaling industry, with its heavy capital investment, has been unwilling – or unable – to reach agreement on adequate regulations or to enforce such as have been reached. Short term, short-sighted exploitation has endangered the long term prospects of the industry – not to mention the whales.

4. Catch controls have consistently been based on the most optimistic figures of whale populations when the only responsible procedure would have been to err on the side of caution. If the size of whale catches cannot be brought under effective control because the policies of the IWC are dominated by the whaling industry, then the IWC should become directly responsible to the United Nations for the restoration of the great whales.

5. Relatively little is known about the social habits of whales, or of their ability to communicate seemingly complex messages. Such studies as have been made indicate that some cetaceans may possess a highly developed intelligence. It would appear to be scientifically prudent and ethically sound to learn much more about whales before permitting any further reduction of the populations. In the past scientists have concentrated their studies on dead whales and some are concerned that their work will be brought to a halt by a moratorium.
Scientific study of living whales is still in its infancy, but may well prove more about population structures and migrations than has been possible from the industrial catches. A moratorium could greatly stimulate such research. It will still be important to monitor whale populations as indicators of the wellbeing of the oceans and their ecosystems. We believe this can be done without killing them.

6. The method of killing whales can only be described as barbaric. The barbed, 160 lb. harpoon explodes inside the whale. Dying frequently takes up to half an hour and can take two hours.

7. We do not believe man any longer needs to hunt whales. We think that in the name of human dignity, the killing should stop, at least until whale products are shown to be essential to human survival and a humane capture technique has been devised.

We the undersigned, belonging to many international conservation, natural history and scientific bodies urge all the delegates to the IWC to vote for a 10 year moratorium on all commercial wha

HRH Prince Bernhard of the Netherlands, President, World Wildlife Fund International
HRH The Duke of Edinburgh KG, KT, President, World Wildlife Fund British National Appe President, Australian Conservation Foundation.
Professor D. J. Kuenen, President, International Union for the Conservation of Nature and Natural Resources, Netherlands
Laurence I. Moss, President, Sierra Club
William A. Nierenberg, Director, Scripps Institution of Oceanography
Cleveland Amory, President, Fund for Animals
John Aspinall
Dr. Gerardo Budowski PhD
Commander Jacques Cousteau
Prof. Jean Dorst
Prof. Rene Dubos PhD
Dr. Paul Ehrlich
Sir Frank Fraser Darling FRSE
Dr. Thor Heyerdahl PhD
Sir Julian Huxley FRS
Prof. Claude Levi-Strauss
Dr. Konrad Lorenz
Dr. Sicco Mansholt
Sir Peter Scott CBE
Dr. J. E. Smith CBE, ScD, FRS
Animal Defence Society Limited
The Conservation Society
The Fauna Preservation Society
Friends of the Earth International
The International Society for the Protection of Animals
Project Jonah
The Royal Society for the Prevention of Cruelty to Animals
The Universities Federation for Animal Welfa
The World Wildlife Fund, British National Ap

(Advertisement in "The Times", Monday 25th June, 1973)

A motion in 1976, supporting both the moratorium and a ban on all whale imports, was signed by over 260 MPs.

Before March 1973, Britain imported a considerable amount of whale meat both from IWC and non-IWC countries (see Table 13). This meat was used by British pet food manufacturers for tinned and unprepared pet food. After considerable public indignation following a BBC Horizon film by Simon Campbell-Jones called "Whales, Dolphins and Men", the Pet Food Association announced that they would introduce a voluntary ban on the use of whale meat in their pet foods. But like most voluntary bans, they are of little use if they do not have the full support of the whole industry.

Then, on 15th March 1973, after mounting public pressure for a ban on whale imports, the Government took positive action. In reply to Parliamentary Questions by David James MP and Sir John Langford-Holt MP in the House of Commons, Anthony Stodart MP, then Minister of State for Agriculture, Fisheries and Food, said "Investigations undertaken by the Department of Trade and Industry into Sperm oil usage have indicated that substitutes for some of its uses are available but that certain industries would face difficulties if it were not imported. Sperm whales have not been over-exploited, but baleen whales have. So to supplement the measures of the IWC, the Government are banning all imports of whale products, except Sperm oil, Spermaceti wax, ambergris and those products incorporated abroad into manufactured goods. The ban will start from midnight tonight."

The statement made solemnly to the House of Commons was, however, untrue. It was later admitted that whales teeth and bones and skin were not then restricted, and more examples regularly come to light.

This decision to ban baleen whale imports was largely a political one. Although the Government had supported the moratorium, its chief scientific advisor, Dr. Ray Gambell of the Institute of Oceanographic Sciences, had not. Dr. Gambell still believes very strongly that whaling should continue on a sustainable yield basis.

The ban, therefore, was due to a combination of public pressure and a fear on the part of the British Government for the future of the seriously depleted Fin Whale stocks. The Government gave the following reasons why the ban did not include foreign manufactured goods:

1) Exporters are not required to declare a whale product content. (It was not explained why this could not be required.)

2) In many cases there would be no valid test to detect such a content. In any case, the quantity of baleen whale products thought to be imported in this form was very small.

When announcing the ban, Mr. Stodart said that the Sperm Whale stocks had not been over-exploited. However, new data **has** recently come to light which shows that there is certainly cause for concern for the Sperm Whales (see page 95).

Yet in August 1974 the Government was still maintaining that " . . . the situation on whale stocks is by no means as serious as you (and others) have been led to believe" (letter from S. F. Wright on behalf of the Prime Minister 28th August 1974). In September K. M. Bromley of the Ministry of Agriculture, Fisheries and Food informed us that, "Sperm whale products were not included in this ban (on baleen whale products) because there is no conservation reason at present to do so". The fact that there is a considerable scientific controversy on the subject of whale populations must raise the question as to whether or not the British Government has "the best scientific information available".

b)   POLICIES

These are the actions we think the Government should be taking:

1)   The British Government is in full support of the 10 year moratorium and therefore, to be consistent with this pledge and the vote at Stockholm, should impose a ban on the import of **all** whale products. Moreover, if the Government forbade the import of all whale products it would give us a position of strength from which to exert pressure on other countries. Some civil servants think that unilateral bans are not effective, but the British Government has imposed import bans on animal products such as vicuna hair, tiger skins and baleen whale meat. In 1970-72, Russia exported 33% and Japan 32% of their Sperm Whale products, at values of $16 m and $8 m respectively. This foreign earnings product is undoubtedly the reason why whaling industries in many countries have been able to keep a hold over their respective governments. Therefore, import bans by individual countries would be bound to have an effect.

Table 12 – Production and Export of Sperm Oil by USSR and Japan.

Japan	1970	1971	1972	Total
Production 000 tonnes	37.5	31.7	25.9	95.1
Export 000 tonnes	10.8	17.7	2.1	30.6
Export $000	2 455	5 500	702	8 657
USSR				
Production·000 tonnes	80.2	68.3	45.8	194.3
Export 000 tonnes	34.1	14.2	16	64.3
Export $000	8 683	3 777	3 606	16 066

Source: FAO Statistics.

(The enormous drop in Japanese exports in 1972 (and high level in 1971) is partly explained by the US ban being introduced. In 1971 many US importers stockpiled.)

Table 13 – Imports of Whale Products into United Kingdom
(weight in metric tons)

	1966	1967	1968	1969	1970	1971	1972	1973	1974
Fresh, chilled or frozen whale meat	17 120	18 330	22 140	14 430	19 200	15 520	8 870	40	–
Sperm oil	13 920	18 820	28 250	16 280	14 110	11 300	7 610	6 430	8 144
Whale oil	4 430	4 470	10 250	2 220	1 520	800	870	420	50.9

The most important reason for wanting a ban on all whale imports is to protect the Sperm Whale – the most ecologically successful of all whale species. Most whaling scientists have been complacent about Sperm Whale populations as they do not appear to have suffered a drastic numerical decline, but it has declined considerably in size and weight (see page 96). As a consequence, the whalers are tempted to catch twice as many Sperm Whales to make up the catch.

The sperm oil currently imported into Great Britain (see table 13) is mainly used as an industrial lubricant. The importer and sole refiner of unrefined sperm oil in Great Britain is Highgate & Job of Paisley and Liverpool. The majority of this oil comes from Australia (3,000 tonnes worth £306,976 in 1973), but a significant amount of this oil also comes from the Netherlands (980 tonnes in 1973). This is because the Netherlands is the centre of Europe's sperm oil market – although there are no Dutch whaling boats left. Highgate & Job exports refined sperm oil to a large number of countries (see page 130). Some firms, including Highgate & Job, have told the Department of Industry that they would suffer if a sperm oil ban was introduced. A solution to this problem might be to set up a licensing system so that companies which have a real need for sperm oil would be allowed small quantities for a limited time. At the same time, the Government should support research into substitutes. (The Government could also give a subsidy to firms which will not use substitutes because they are marginally more expensive.)

2)   The British Government should also initiate import bans and encourage overt support for the 10 year moratorium en bloc within the EEC. (Trade in whale products by EEC countries is shown on page *133.)*

3)   Friends of the Earth think that the British Government should take much more positive and overt action outside the IWC to put pressure on Japan and the USSR to stop whaling. Mere lip service to the moratorium will change

nothing. Britain has a special duty to ensure effective international action because of her part in the decimation of the Blue and Fin populations in the Antarctic.

This pressure could be made in a variety of ways. One of the most effective methods might be the one which the Americans have already adopted — the threat of economic sanctions if Japan and the USSR do not comply with international regulations on marine resources. (see p. 17)

4)   China, South Korea, Peru, Chile, Spain and Portugal are countries which hunt whales, but do not belong to the IWC or adhere to any international regulations. The Government should influence countries with which we have close political, economic and traditional ties to stop whaling and/or join the Commission.

5)   The Government should act within the United Nations, especially UNEP and the FAO and fisheries commissions of which it is a member to take the IWC under its auspices.

6)   The Government should act through international organisations (such as the Antarctic Treaty) to ensure that the living resources of the Antarctic (i.e. krill and restored whale stocks) are harvested on an environmentally sound basis and that their products are distributed equitably.

APPENDIX I

WHALES IN ANCIENT HISTORY & MYTHOLOGY

Early societies which knew the whales tend to embody in their folklore great respect for these creatures. In some cases cetaceans seem to have been held in such esteem that, rather than profit from the death of a stranded whale, it was the duty of the tribe to do everything in its power to help the poor beast on its way. Failing that, the death of a whale was the cause of sorrow.

For instance, in Vietnamese culture (which is itself as close to extinction as any of the great whales), whales are regarded as the knight errants of the God of the Waters sent to help fishermen in trouble and to carry shipwrecked sailors to safety. When a dead whale or dolphin is found, its finder goes into mourning for a period of three months, after which time the bones of the whale are transported to a sanctuary ranking as equal in importance to the tombs of royalty.

The most famous of all mythological cetaceans is Jonah's whale. In fact the original Hebrew word tannin means practically any kind of monster and the story is usually considered to be allegorical. However, in the New Testament the word ketos (cetacean) is used, and even more interesting is the use of the word anbar in the Ethiopic Bible. This word is derived from the Arabic word for ambergris, and identifies the whale clearly with the Sperm Whale. Anbar, or amber as it became known in Europe, originally applied to the highly prized product of the Sperm Whale but later became used to describe fossil resin used in jewellery making, which was also found cast up on beaches. The whale product was then distinguished under the same name — amber gris, or grey amber.

In European culture, too, cetaceans were held in high esteem. In ancient Greece it was a serious crime to take the life of a dolphin, these 'rulers of the waters' being treated with the same respect as any rulers of the land. Admittedly different communities place a different aesthetic value on whales. For example, the Eskimos and other aboriginal groups use the whale as a lifeline in their otherwise harsh environment, but this in no way prevents their appreciation of the whale's intrinsic value. These societies see the whale in context and this is shown in their art, mythology and folklore.

The Eskimo Myths

The Koryak referred to the whale in their mythology, "The Founding Myth", which explains creation. The myth centres around the Indian, 'Big Raven', who one day came upon a stranded whale and decided to help it back into the ocean. Realising that it was too big, he asked 'Existence' how he could save the creature when he was not strong enough to lift it. 'Existence' replied and told him to go to an area of the forest where moonlight hits the ground and there he would find mushrooms to eat. After Big Raven had eaten the mushrooms, Existence asked him if he now felt strong enough to set the whale on its way. He did.

Norse Myths

Thirteenth century Icelandic folklore is riddled with stories of whales, both good and bad. Right Whales and Fin Whales were not only good to eat but were extremely clean-living, feeding only on the 'darkness and rain that fall onto the sea'. They could also be relied on to help protect ships from the murderous advances of the evil whales — horse whales, pig whales and red whales — mention of which was taboo. Icelandic and Norwegian fishermen were convinced that as long as men did not fight amongst themselves, the good whales would every year drive shoals of herring toward the shore and into their nets. It was their belief, which survived into the second half of the nineteenth century and caused great opposition to the introduction of modern whaling methods in the North Atlantic, that caused the introduction of laws controlling whaling in 1904.

Royal "Fish"

In England, Wales and Ireland all species of whales, together with sturgeon, became the property of the monarch after a law was passed during the reign of Edward II. At that time, England's jurisdiction extended well south of the Channel into the part of continental Europe now known as France, and where whaling was already being practised; it is not such a useless prerogative as it might at first seem. The act became used to acquire stranded specimens for the national collections housed in the Natural History Museum in South Kensington, London. HM Customs and the Receivers of Wrecks notified the museums of any whales stranded who collected the victims for dissection and the preparation of specimens.

APPENDIX II

THE HISTORY OF WHALING

a)   THE HISTORY OF WESTERN WHALING

From Spears to Warheads

The history of whaling falls into three main phases: early or European whaling, American Sperm whaling and pelagic or modern whaling. They all have similar patterns in that the industry has always declined because of over-fishing.

The earliest records of European whaling date from the ninth century, although small whales were certainly hunted since Stone Age times. The Basques were the first people to make an industry out of whaling. They hunted the Biscayan Right Whale which migrated along their coasts. The Right Whales were so called as they were the 'right' whales to catch because they were slow swimmers and floated when dead. Thus they were not too hard a match for the Basques who pursued them in open boats and killed them with hand-held harpoons. Whaling was profitable to the Basques even though by today's standards little of the whale was utilised. They boiled down the blubber and used the oil for their lamps and for heating. (The oil from baleen whales has become known as train oil, not because it has anything to do with trains; it comes from the Dutch word 'traan' — a tear drop.) The meat was eaten if it was fresh enough.

The Biscayan Right Whales became scarce in their coastal waters (and they have never returned or recovered from the onslaught); the Basques built stronger boats and searched further afield for them. They went to Iceland and Greenland and by 1538 as far away as Newfoundland. During the early 1600s the English and the Dutch picked up the Basque whaling techniques and began whaling along the coasts of Jan Mayen and Spitzbergen. There was great rivalry between the English and Dutch whalers but it was the Dutch who made the improvements in whaling techniques. They developed the wooden slipway and also experimented with rendering down parts of the whale other than the blubber. The whalers' main catch was another kind of Right Whale — the 60 foot Greenland Right (Bowhead) — one of the most bizarre looking of all the species of whale with its huge mouth shaped like an upside-down half-moon. These whales were particularly valuable for their whale bone which was used in many kinds of fashionable clothing, such as bodices, bonnets and bustles. At this time, whale bone fetched £2 250 a ton and each whale yielded about 1½ tons of whale bone and 25 tons of oil. As the whale bone was nearly 14 times more valuable than the oil much of the oil and the remainder of the whale was wasted. After 1630 Greenland Right Whales had become hard to find around the bays of Spitzbergen, so the whalers had to go out to sea to find them. By 1720 they were searching around the west coast of Greenland and Baffin Land, and by 1770 the whales had literally disappeared from these areas too, so the Dutch eventually turned their attention to another species — seals.

The American War of Independence gave a boost to the declining British whaling industry as Britain could no longer rely on her colonies to provide her with whale oil; so she had to get it herself. The British Government granted a subsidy (bounty) of 20s per ton to all whaling ships of 200 tons or more. (This was later increased to 40s per ton.) Hull and Dundee became the chief whaling ports and by 1788 there were nearly 250 English and Scottish whaling ships in Northern waters. The heyday of British Arctic whaling lasted from 1780 to the early nineteenth century. As there were no regulations on the amount, sex or size of the whales caught, it was inevitable that the industry would soon decline.

American Sperm Whaling

The second phase of whaling began along the coasts of North America. The Indians had whaled in their open boats long before the early American whalers started to catch Southern Right, Humpback and the California Grey Whales — all species which swam near the shores on their migratory routes (or in the case of the Grey Whale, bred in the coastal lagoons). As with the Basques and Europeans whalers before them, they soon had to venture further out to sea as the whales became scarce near the coast. On one of these expeditions in 1812 the first Sperm Whale was killed by chance. The American Sperm Whale industry began to flourish and whaling stations sprang up at Nantucket, Mystic and New Bedford. Fortunes were made from the valuable sperm oil and spermaceti, a waxy substance from the head (case) of the Sperm Whale used for lighting and heating, despite the activities of the numerous pirates who preyed on the small whaling ships. It was this period of whaling which Herman Melville so graphically portrayed in his book "Moby Dick".

The American Sperm Whale industry suffered a series of setbacks such as the French Revolution (which meant that markets for Sperm oil in France and England were closed), and the war of 1812 (in which most of her whaling fleet was destroyed). However, the demand of whale oil led to the quick revival of her fleet, and from 1835 to 1846 the industry enjoyed its most prosperous decade. In 1842 there were nearly 600 American whaling ships in operation. The gold rush which drew people away from the whaling industry, and a financial slump in 1857 sent the price of oil plummeting down and were setbacks to the industry. The final blow, however, was the discovery of petroleum which superseded Sperm oil as a fuel and an illuminant.

Modern Whaling

Two new inventions revolutionised the industry and heralded the era of modern whaling. These were the explosive harpoon gun, designed by the Norwegian, Svend Foyn, in 1868, and steam power. These two inventions meant that attention could now be turned to the hitherto unexploited faster and larger whales, the rorquals, the Blue, Humpback, Fin and Sei.

The Norwegians were the first people to establish themselves in the Antarctic.

(Carl Anton Larsen first visited the Antarctic to kill seals but realised that whaling was a possibility and he later returned to set up the first land station in Grytviken, South Georgia.) At first, land stations were set up at South Georgia and the Falkland Islands, but in 1908 the British Government (who governed these Dependencies) became worried about the effects of uncontrolled whaling on the whale populations. So they introduced a licensing system and also stipulated that all parts of the whale must be utilised, in order to stop the then extremely wasteful practice of just using the blubber. They also stipulated that mothers and calves should be protected and later levied a tax on each barrel of oil that was processed in the Dependencies. These well-intentioned controls were deeply resented by the Norwegian whalers who set about inventing a way whereby they did not have to have a land base. This led to the development of the modern floating factory ship, a huge vessel into which a whale could be hauled up the slipway and processed while its attendant smaller and faster catcher ships could chase down the whales. The invention of the factory ship meant that the whalers could pursue whales in open waters and stay at sea for months on end without having to return to base.

The first of this type of ship to operate in the Antarctic was the "Lancing" in 1925. Five years later 38 factory ships were whaling in the Antarctic and during the 1930-31 season, nearly 43 thousand whales were killed, and over 29 thousand of these were Blue Whales. The slaughter continued until the outbreak of the Second World War when man used his technology to kill another species — himself.

During the 1930s the industry realised they would have to impose some curbs on themselves if whaling was to continue on a long-term basis. The League of Nations was the first organisation which tried to give some protection to the whales, but its attempts to do so in 1924 and 1927 failed. An agreement was reached in 1931 and subsequently revised in 1937, 1938 and 1939, but it was not until 1944 that an excessively-high quota of 16 000 Blue Whale units for all waters south of 40° South latitude was agreed upon.[1] This catch limit was later incorporated into the International Convention for the Regulation of Whaling which was signed by Argentina, Australia, Brazil, Canada, Chile, Denmark, France, the Netherlands, New Zealand, Norway, Peru, USSR, UK and the USA in 1946, and which entered into force in 1949. The purpose of the Convention was to safeguard for future generations "the great natural resources represented by the whale stocks" and "to protect all species of whales from further overfishing". The Schedule to the Convention gave protection to the Grey and Right Whales and females with calves. The Convention established the International Whaling Commission.

The world shortage of oils and fats meant that whaling was bound to start again immediately after the war, and it is this period of commercial whaling that is the most damnable. The BWu as a quota system meant that the whalers went after the largest whales, as catching one Blue Whale involved far less effort than catching six Sei. However, the Japanese used the BWu to their advantage. They were much

(1) More realistic quotas were not set until 1965.

more interested in whale meat (especially Sei whale meat), rather than whale oil, as they had a tradition of eating whale meat and home markets for this were already established. So it was to their advantage to kill 6 Sei whales (although by weight 4 Sei = 1 Blue Whale) even though their oil yield was poor. They were able to maintain their industry in spite of declining catches of Blue and Fin Whales and keep within the quotas based on the BWu. The BWu was therefore not abolished until 1972 even though the Special Committee of Three Scientists recommended to the IWC that it should be abolished in 1963.

As the BWu did not afford protection to individual whale species, the stocks of rarer whales were given no particular protection. The preferred species were the largest, so the Blue, Fin, Humpback and Sei were decimated in turn by the Norwegians, South Africans, British, Japanese, Russians and Dutch. The whalers then turned their attention to whale species which were not included in the BWu – the Sperm and the Minke.[2]

b)   THE HISTORY OF JAPANESE WHALING

The Japanese have been whaling off the coast of Japan for more than a thousand years, but it was not until the early seventeenth century that commercial whaling was established in Taichi in the Wakayama Prefecture. It soon spread throughout the islands and by the early nineteenth century there were 30 coastal whaling bases. There are many old prints and paintings of Japanese people in small fishing boats catching whales with nets and harpoons. There are also paintings of baleen whales being dismembered on the beaches. These paintings were mainly made during the Edo-Tokugawa period of 1600-1853. A Japanese proverb of the time ran, "when one whale is caught it makes seven villages prosperous".

Hand harpoons were first used to catch whales, but in the late seventeenth century they seemed to have changed to net catching. This method required the use of about 12 boats. The Japanese changed their methods after seeing the Russian Pacific Whaling Company in the Japan Sea in 1898 which used modernised Norwegian methods and equipment and which supplied them with 1 000 tons of whale meat with no difficulty.

The Japanese had the same advantage as the Basques in that they could catch whales which migrated past their shores, e.g. Humpback and Grey Whales. However, comparatively few whales were taken. From 1698 to 1878 an average of 13 whales were caught each year. Despite the low level of the Japanese catch, by the nineteenth century whales began to be scarce in Japanese waters. This could have been due to the influence of the American whalers of which there were 300 in Japanese waters in the mid-nineteenth century. In 1899 the Choshu-mauru, the first Japanese steel catcher boat, operated off Korea and killed 155 whales. By 1919 the Japanese Government had to restrict the number of catchers in Japanese waters to 30 for conservation reasons.

---

(2) The trend towards exploiting smaller and smaller species of whales and dolphins continues.

Japan did not start whaling in the Antarctic until 1934, after Nippon Suisan had bought a factory ship and five catcher boats from the Norwegians. Taiyo Fishery Company and the other companies followed suit. By 1939 Japan had acquired six factory ships, all of which were destroyed during World War II. In 1951, the Allied Powers founded the Japan Development Bank to help finance the reconstruction of the fishing industry (which included whaling), Japan continued to buy up the whaling ships of other nations such as Netherlands and United Kingdom who had discontinued whaling. With their boats, they also inherited their share of the overall quota. (The IWC whaling nations decide amongst themselves what the national shares should be. If a country stopped whaling and sold its whaling ships to another whaling country, that country assumed it inherited the other country's quota, but the USSR vigorously contested such an assumption.)

APPENDIX III

CHECKLIST OF THE LIVING CETACEANS

Sub-Order MYSTICETI
    Family BALAENIDAE

Balaena glacialis	Black Right Whale
mysticetus	Greenland Right Whale
Caperea marginata	Pygmy Right Whale

    Family ESCHRICHTIIDAE

Eschrichtius robustus	Grey Whale

    Family BALAENOPTERIDAE

Balaenoptera acutorostrata	Minke Whale
edeni	Bryde's Whale
borealis	Sei Whale
physalus	Fin Whale
musculus	Blue Whale
Megaptera novaeangliae	Humpback Whale

Sub-Order ODONTOCETI
    Family PLATANISTIDAE

Platanista gangetica	Susu
Inia geoffrensis	Bouto
Lipotes vexillifer	White Flag Dolphin
Pontoporia blainvillei	La Plata Dolphin

    Family DELPHINIDAE

Steno bredanensis	Rough-toothed Dolphin
Sousa teuszi	
plumbea	Plumbeous Dolphin
lentiginosa	Speckled Dolphin
borneensis	Bornean White Dolphin
chinensis	Chinese White Dolphin
Sotalia fluviatilis	Tookashee
guianensis	Guiana Dolphin
Tursiops truncatus gilli	Bottle-nosed Dolphin
Grampus griseus	Risso's Dolphin
Lagenorhynchus albirostris	White-beaked Dolphin
acutus	Atlantic White-sided Dolphin
obliquidens	North Pacific White-sided Dolphin
australis	Black-chinned Dolphin
cruciger	Hourglass Dolphin
obscurus	Dusky Dolphin
Lagenodelphis hosei	Sarawak Dolphin
Stenella longirostris	Eastern Pacific Spinner Dolphin
roseiventris	Hawaiian Spinner Dolphin
dubia	Spotted Dolphin
caeruleoalba	Striped Dolphin

Delphinus delphis	Common Dolphin
Lissodelphis borealis	Northern Right-whale Dolphin
peroni	Southern Right-whale Dolphin
Cephalorhynchus commersoni	Commerson's Dolphin
eutropia	White-bellied Dolphin
heavisidei	Haviside Dolphin
hectori	Hector's Dolphin
Peponocephala electra	Broad-beaked Dolphin
Feresa attenuata	Pygmy Killer Whale
Pseudorca crassidens	False Killer Whale
Globicephala melaena	Common Pilot Whale
macrorhyncha	Short-finned Pilot Whale
Orcinus orca	Killer Whale
Orcaella brevirostris	Irrawaddy Dolphin
Phocoena phocoena	Common Porpoise
sinus	Gulf of California Porpoise
dioptrica	Spectacled Dolphin
spinipinnis	Black Porpoise
Neophocaena phocaenoides	Black Finless Dolphin
Phocoenoides dalli	Dall's Porpoise

Family MONODONTIDAE

Delphinapterus leucas	White Whale
Monodon monoceros	Narwhal

Family PHYSETERIDEA

Physeter catodon	Sperm Whale
Kogia breviceps	Pygmy Sperm Whale
simus	Dwarf Sperm Whale

Family *ZIPHIIDAE*

Tasmacetus shepherdi	Tasman Beaked Whale
Mesoplodon bidens	North Sea Beaked Whale
europaeus	Antillean Beaked Whale
mirus	True's Beaked Whale
pacificus	Longman's Beaked Whale
grayi	Scamper-down Whale
hectori	Hector's Beaked Whale
stejnegeri	Bering Sea Beaked Whale
carlhubbsi	Arch-beaked Whale
bowdoini	
ginkgodens	Ginkgo-toothed Whale
layardi	Strap-toothed Whale
densirostris	Dense-beaked Whale
Berardius arnouxi	Southern Giant Bottle-nosed Whale
bairdi	North Pacific Bottle-nosed Whale
Hyperoodon ampullatus	Northern Atlantic Bottle-nosed Whale
planifrons	Flat-headed Bottle-nosed Whale
Ziphius cavirostris	Goose-beaked Whale or Cuvier's Beaked Whale

Based on "A List of the Marine Mammals of the World" by Dale W. Rice and Victor B. Scheffer (1968), US Fish & Wildlife Service Special Scientific Report — Fisheries No. 579.

APPENDIX IV

CURRENTLY EXPLOITED SPECIES

a)   SPERM WHALE (or CACHALOT) (*Physeter catadon*)

The Sperm Whale is one of the most intriguing species in the order Cetacea. It is the largest of the toothed whales, the males growing to 60 feet or more. Unlike the baleen whales, the females are smaller than the males — about half the size. A deep diving species, it can swim at a depth of 2,000 metres for prolonged periods. It feeds principally on squid and other cephalopods. It is quite different in appearance from all the other great whales, having an enormous head, with small, toothed lower jaws. The enormous forehead contains the valuable spermaceti wax. The Sperm Whale is gregarious and polygamous. They have a strong schooling instinct and the females, together with their offspring, remain in warm waters in harems dominated by a single bull. They occur in practically all the seas and oceans but the non-breeding males migrate periodically to the polar regions, and it is these non-breeding males which form the basis of the Antarctic Sperm Whale industry.

Sperm Whaling has had a long and rather turbulent history, and has probably suffered more ups and downs due to fluctuations in the market prices of oil, than the baleen whaling industry. During the 18th and early part of the 19th century it was almost an American monopoly centred on the ports of Nantucket and New Bedford. The American War of Independence caused a major slump in the industry and Britain systematically destroyed whaling boats and created tariff walls. Similar destruction occurred during the American Civil War, but by that time the demand for sperm oil was declining with the introduction of kerosene. The centre of whaling then shifted from New England to the West Coast, particularly San Francisco. At its peak around 1846, 735 ships set out, but from 1877-86 there was only an average of 159. The American Sperm whaling capture techniques still survive in the open boat whaling practised in the Azores. The long-term effects of the American whaling industry on the Sperm Whale populations, whilst by no means insignificant, at least locally, were certainly small by today's standards. It was not until the advent of modern techniques that serious inroads were made into the populations and that has really only occurred in the last couple of decades. (Sperm Whales have been the main target of the commercial pelagic whalers in the Antarctic and North Pacific since the early 1960's and now constitute over 40% by weight of the total catch of all species of whales.)

Until very recently whaling scientists, with few exceptions, have tended to be complacent about the state of the stocks of Sperm Whales, assuming that they were well above msy levels, and that the present rate of cropping was not having any significantly bad effects on the stocks. However, from time to time, scientists have pointed out that perhaps Sperm Whales ought to be treated rather differently from the baleen whales as their population structure and breeding biology are radically different (i.e. they are polygamous, herd animals and slow breeders, having a single young only once every *4 to 5* years). Even more recently it has been pointed out that there have been marked, though not immediately obvious, changes in the catch of Sperm Whales. It is this latter trend which gives particular grave cause for concern and which we believe should be investigated thoroughly before we can be sure what the msy is, but meanwhile we must be very conservative.

The Sperm Whale catch is regulated by quotas set separately by sex and by stocks in the southern hemisphere. At the 1974 IWC meeting, the Scientific Committee said that all Sperm Whale quotas should be set by stocks. However, Governments resisted this advice for the same reasons that they resisted constraints on the Blue Whale catch — because it would be more expensive.

The setting of quotas separately for male and female Sperm Whales had proved very profitable to the whaling industry. It was originally argued that as male Sperm Whales could easily be identified because of their larger size, and that as they are a polygamous species, the bulls in the Antarctic could be harvested at a relatively high level without harming the overall breeding potential of the species. The Scientific Committee then recommended a high quota for female Sperm Whales in order to "bring the sexes into a better balance" — an act of doubtful and illogical wisdom. The delegation from the USSR suggested higher size limits for Sperm Whales at the 1974 IWC meeting. Obviously they had had second thoughts. This suggestion is all the more significant coming from the Russian delegation as most of their catch is of Sperm Whales.

If, however, the weight (which reflects the total living mass (biomass) and the age of the individuals within a population) is considered, then marked changes in the present populations can be seen when compared with those of previous years.

Figure 4 shows the change in weight of males in the Antarctic as calculated from the length measurements recorded in the International Whaling Statistics. The decline in average length from about 54 to around 45 feet is quite dramatic, but the decline in average weight is in the order of 47 tons to 27 tons — an obviously disastrous decline. Cousteau, in his book, "The Whale" (1972), records that in the days of Moby Dick, Sperm Whales were often 90 feet long and Bullen talks of 70 foot Sperm Whales in the 1890's. Similar, though not always so obvious, declines are known to have occurred in most other Sperm Whale populations. A comparison of the lengths of animals caught in 1932 with those 40 years later, shows that this decline is not just due to the fact that the larger bulls have been removed from the population, but in fact the whole range of sizes found within the Antarctic population of male Sperm Whales has shifted downwards. This can certainly be explained to a large extent by saying that now the larger bulls are no longer available, the whalers are seeking smaller whales; but it still seems likely from the size distribution curve that there has been an overall decline in the average size of the whales in the population. It has also been suggested (Dr. P. E. Purves, personal communication) that the effects on the gene pools of Sperm Whales ought to be considered: what happens to the genetic structure of a population if all the larger animals (regardless of age) are killed? It is quite possible that in a relatively short period such intensive selection could produce populations of smaller Sperm Whales.

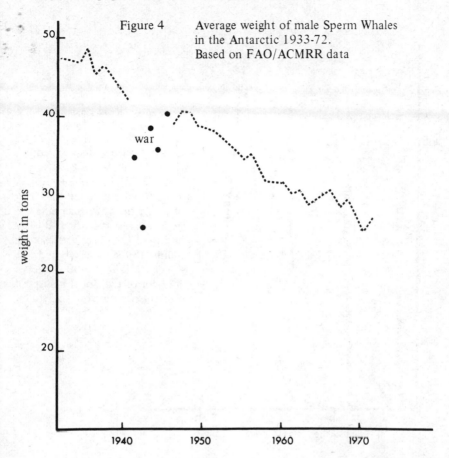

Figure 4   Average weight of male Sperm Whales
in the Antarctic 1933-72.
Based on FAO/ACMRR data

weight in tons

war

Figure 5

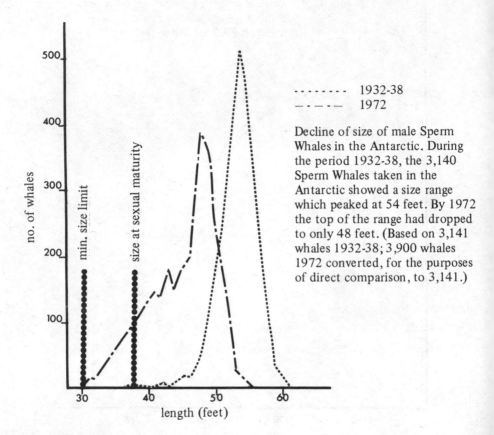

Decline of size of male Sperm Whales in the Antarctic. During the period 1932-38, the 3,140 Sperm Whales taken in the Antarctic showed a size range which peaked at 54 feet. By 1972 the top of the range had dropped to only 48 feet. (Based on 3,141 whales 1932-38; 3,900 whales 1972 converted, for the purposes of direct comparison, to 3,141.)

A further and equally serious problem with Sperm Whales is that the size limits prescribed by the IWC are such that they may be hunted before they have reached anything like maximum size. By contrast, the minimum size at which baleen whales may be hunted is much nearer the mean maximum size. This means that Sperm Whaling is inefficient in as much as if the whales were left in the sea longer, they would continue to grow at a fairly fast rate, producing a greater yield if they were killed at a larger size, i.e. three quarters of the potential growth rate of male Sperm Whales occurs after they are available for exploitation.

It may be that *some* stocks of Sperm Whales are at msy level by numbers. However, the stocks in biomass terms are very much below the accepted msy.

With such reduced populations the distinctive breeding biology of the Sperm Whales could become disrupted, although in what way is still unclear. It is thought by some, for example, that growth and reproductive rates of whales are density

98

dependent (see page 66) and in the case of the female Sperm Whale it is suggested that ovulation is induced by the presence of a male; how these and other factors are affected by a general reduction of population levels is unknown.

Estimates of Sperm Whale populations using CPUE* data are admitted to be in any case very difficult with present data. However, to make matters worse, there have been big changes in the effectiveness of catches in recent years as a consequence of the use of spotter aircraft and ASDIC to track submerged Sperm Whales. This *was* done in South Africa and it is not clear to what extent the method is used by the Russians and Japanese. The IWC has only recently looked into this. It could have the effect of changing the apparent rate of decline of Sperm Whales several-fold.

Since the arguments in favour of a continued commercial whaling are almost entirely based on short-term economics, it is perhaps worth pointing out that with the decline in the biomass of the Sperm Whale populations there has also been a drastic decline in the productivity of Sperm Whaling.

In 1974 Ray Gambell estimated the size (but only in numbers) of the Sperm Whale populations (see table 14). It can be seen from this that even disregarding the weight factor discussed above, that the quotas set are far higher than the numbers that could actually be caught in the previous year.

Table 14 – SPERM WHALE DATA (NUMERICAL)

| | Southern Hemisphere | | North Pacific | |
	Male	*Female*	*Male*	*Female*
Estimated original population	ca. 257,000	ca. 330,000	ca. 195,000	ca. 152,000
Estimated population *in 1974*	ca. 128,000	ca. 259,000	ca. 91,000	ca. 125,000
1973/74 quota	8,000	5,000	6,000	4,000
1973/74 catch	6,709	4,183	*4,419*	*3,708*
1974/75 quota	8,000	5,000	6,000	4,000
1974/75 catch	*7,097*	*4,737*	*4,261*	*3,598*
1975/76 quota	*5,870*	*4,870*	*5,200*	*3,100*
1975/76 catch	*4,022*	*3,024*	*4,261*	*3,610*
1976/77 quota	*3,894*	*897*	*4,320*	*2,880*
1976/77 catch	*3,308*	*767*		
1977/78 quota	*4,538*	*1,370*	*0*	*763*
1977/78 quota revision			*5,105*	*1,339*

After data supplied by Dr. R. Gambell and others

---

*see p. 63

At the June 1977 meeting of the IWC, North Pacific quotas for Sperm Whales were cut to virtually zero, but their Special Commission Meeting of December 1977 raised this limit to 6,444. For a catch that is intended to be sustainable, there can surely be no clearer indication of the IWC's inability to "manage" this cetacean.

b) MINKE WHALE (or LESSER RORQUAL) (*Balaenoptera acutorostrata*)

The Minke Whale (named after a Norwegian whaler, Miencke) is the smallest of the rorquals, growing to just over 30 feet. It has characteristic yellowish-white baleen plates, and often has a pale patch on the flipper.

The little information regarding this species has mainly come from sightings by Japanese whalers. But as the FAO/ACMRR paper pointed out, there is marked sexual and age segregation, and its distribution is patchy. Sighting is made difficult by the fact that it is often only visible at a distance when it leaps; otherwise it often lies hidden in the wave troughs. In some places it leaps frequently, in others not at all.

This species has long been one of the most important 'small' whales. This is primarily because it is one of the largest whales that can be taken by the 'small' whale coastal fisheries. But, with the decline of the great whales, the commercial pelagic whaling fleets have increasingly turned their attention to this species and it is now the most important baleen whale to the pelagic fleets of the USSR and has contributed to the maintenance of the baleen catch in recent seasons.

In 1964/65 the total world catch of Minke Whales was 2 917, of which only 6 were taken in the Antarctic. By 1972/73 the catch had risen to 5 745 in the Antarctic alone and in 1973, 9 288 were killed. Japan now operates one factory ship and four catchers exclusively for Minke Whales (Minke Whales constituted nearly one third of the total of 190 000 tons taken in the 1973/74 season).

The 1974/75 quota for Antarctic pelagic whaling was set at 7 000 Minke Whales, although Japan and Russia were pressing for at least 9 000. No quotas were set at all for the North Pacific, or for land-based whaling operations. This was in spite of the fact that the advice to the IWC from its Scientific Committee stated that there was virtually no information about the state of Minke Whale populations, and they recommended that the **maximum** yield from the **total world** Minke Whale stocks should not exceed 7 000 individuals. The FAO/ACMRR working party suggested that the present data are "not incompatible with the view that **Minke stocks are even, already at or below msy level**". There is, therefore, considerable disagreement among whaling scientists over the size of Minke Whale populations. It is interesting to note that the Japanese consistently give higher stock estimates and sustainable yields than anyone else.

c) FIN WHALE (*Balaenoptera physalus*)

The Fin Whale is the second largest of the rorquals, growing to some 80 feet.

It has a similar streamlined appearance to the Blue Whale, but is more slender. The lower jaw is white, but only on the right hand side.

Fin Whales are rapidly reaching commercial extinction, but as they are the largest and thus most valuable in terms of whale products, of the currently exploited whales, the whalers still find it profitable to kill them. It is one of the most revealing weaknesses of the IWC that the Fin Whale is not completely protected, as it is universally uncontested that the Fin Whale is far below msy level. In the southern hemisphere Fin Whales are estimated to be at about one fifth of their original population, and in the North Pacific they are at about one third of the original level (see Table 16 *p. 105*).

It has been suggested that the rate at which Fin Whales were exploited in recent years ties in with the argument (outlined on p. 59) that once an animal has declined to a certain level, it then continues to be economical to bring the species concerned to commercial extinction as rapidly as possible. It also continues to be economical if the catch can be made up with other species; this is what has happened with all depleted stocks. The late Dr. N. A. Mackintosh showed in 1965 that (on the estimates available at the time) had Fin Whales been totally protected in 1960, their stocks would have recovered by about 1970 sufficiently to give a yield of 20 000 p.a. By 1968 the recovery time had extended to the late 1980's. Dr. Allen (IWC report 1972/73) has estimated that it will now take the Antarctic stocks of Fin Whales 50-60 years to recover to 90% of the original level.

d)   **SEI WHALE** (*Balaenoptera borealis*) and
      BRYDE'S WHALE (*Balaenoptera edeni*)

The Sei Whale (sei — pronounced sigh — is the Norwegian for the coalfish, a relative of the cod. The Sei Whale appears off the Norwegian coast at the same time each year as the fish), which reaches a length of about 60 feet, is rather plumper in appearance than the Fin and Blue Whales and has a characteristic greyish-white baleen, which is frayed on its inner edge. Bryde's (pronounced Bruders) Whale grows to about 50 feet and, although superficially similar to the Sei, has a smaller dorsal fin and longer, stiffer baleen plates. Bryde's Whale has a wider palate and also two ridges on the top of the head.

Sei Whales have only been fished intensively in the Antarctic and in the North Pacific since the early 1960's when the Fin Whale stocks began to collapse. One factor which must have contributed to their decline was the BWu as the whalers could obtain 50% more product weight in taking each BWu as Sei rather than as Fin Whales (1 Blue = 2 Fin or 2½ Humpback or 6 Sei). Some whaling scientists believe that Sei Whales still exist in relatively large numbers, but the hard evidence for this is elusive.

Dr. Ray Gambell, in 1974, gathered various estimates of the population sizes and suggested that in the southern hemisphere there were 82 000-97 000 and in the

North Pacific there were 34 000-38 000 Sei Whales. The original populations were about 150 000 and between 58 000 and 62 000 respectively. He estimated the msy levels for the two populations at 52 000 and 33 000. Other scientists are not so optimistic. The FAO working party was clearly worried about the health of the Sei Whale populations. They were "puzzled by the apparent reluctance of the IWC Scientific Committee to revise downward its estimates of Sei stock size and yields, in the face of continuing declines". The members òf the working party also thought that there should be a special meeting on the Sei Whale to re-appraise the situation in the hope that this might prevent the Sei Whale from being over-exploited as were the Blue, Fin and Humpback before it. The report continues, "in the case of the Sei Whale sightings data showing continuing drastic decline should probably be given great weight, especially as the different stocks of Sei all seemed to be declining, and these should be managed separately".

Fig. 6 Catch of Sei (and Bryde's) 1932-72

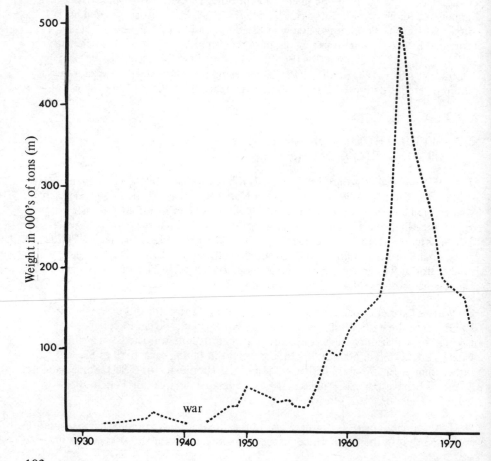

One of the areas in which the Sei Whale has declined most rapidly is in the region around South Africa.

Figure 7 shows the rapid depletion of a small stock of Sei Whales that was fished to commercial extinction within a decade. In fact the FAO group was "puzzled by the decision of the scientific committee not to revise its earlier estimates of the current sustainable yield of Southern Sei Whales, in the face of continuing declines in sightings off South Africa and in CPUE".

It has also been suggested that the effects of the low level fishery to which the Sei Whale was subjected may have far reaching results, which still have to be investigated.

Fig. 7   No. of Sei Whales caught of South Africa 1963-72,
         showing the virtual extinction of a small population

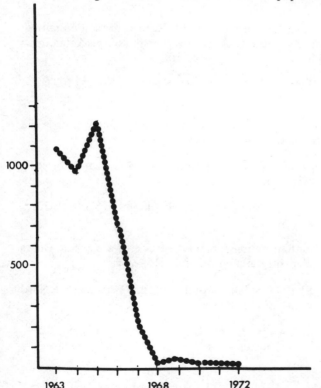

Another worrying aspect of the Sei and Bryde's Whale fishery is that the two species are rarely distinguished by commercial whalers. This is because they are virtually indistinguishable in the water, but once on board the flensing deck there is no reason whatsoever why they cannot be identified. This is quite obviously a source of considerable error. Both the Commission and the Scientific Committee have stressed that quotas should be set both for species and various populations of the individual species. Yet it seems that little effort has been made to set separate quotas for the Sei and Bryde's. The few separate statistics available suggest that the proportion of Bryde's Whales is very significant, at least in the North Pacific. For instance, the Russian catch in the North Pacific in 1970/71 consisted of 638 Bryde's and only 296 Sei. The combined Russian and Japanese catch for the same period reduces the proportion to just ober 30% Bryde's out of a combined total of over 3 000.

Sources to Table 15 and 16

1. International Whaling Commission, 1973, Report of the Scientific Committee. Rep. int. Commn. Whal. 24

2. Ohsumi, S. & Masaki, Y., 1974, Status of whale stocks in the Antarctic, 1972/73. Rep. int. Commn. Whal. 24.

3. Chapman, D.G., Allen, K.R. & Holt, S.J., 1964, Reports of the Committee of Three Scientists on the special scientific investigation of the Antarctic whale stocks. Rep. int. Commn. Whal. 14:32-106.

4. Gulland, J.A., 1972, Future of the blue whale. New Scientist 54 (793): 198-9

5. Ichihara, T. & Doi, T., 1964 Stock Assessment of Pigmy Blue Whales in the Antarctic. Norsk Hvalfangsttid. 54 (6):145-67.

6. International Whaling Commission, 1973, Sperm Whale assessment meeting. Rep. int. Commn. Whal. 23:55-88.

7. Ohsumi, S. & Wada, S., 1974, Status of Whale Stocks in the North Pacific, 1972. Rep. int. Commn. Whal. 24.

8. Wada, S., 1973, The ninth memorandum on the stock assessment of whales in the North Pacific. Rep. int. Commn. Whal. 23:164-9.

9. Ohsumi, S. & Wada, S., 1973, Stock assessment of Blue Whales in the North Pacific. Document IWC/24/SC/13 (mimeo).

10. Rice, D.W. & Wolman, A.A., 1971, The Life History and Ecology of the Gray Whale (Eschrichtius robustus). American Soc. Mammalogists special publ. No. 3.

11. Ohsumi, S. & Fukuda, Y., 1974, Revised Sperm Whale population model and its application to the North Pacific Sperm Whale. Rep. int. Commn. Whal. 24.

12. Talbot, Lee, 1974, History, Status and Conservation of the Great Whale Populations. AGM of American Assoc. for Advanced Science.

Table 15 – Whale Populations in the Southern Hemisphere

Species	Estimated Original Population	Present Population Estimates	% of Original Population surviving (worst estimate)*	Sources	% Variation from mean of present Estimates
Fin	375 000-425 000	82 000- 97 000	19%	1, 2, 12	± 8%
Sei	ca. 150 000	82 000	55%	2, 12	—
Minke	220 000-299 000	150 000-298 000	50%**	1, 2, 12	±33%
Blue	ca. 200 000	5 000- 6 000	2.5%	1, 2, 3, 4, 12	± 9%
Pygmy Blue	ca. 10 000	ca. 5 000- 6 000	50%	1, 2, 4, 5, 12	± 9%
Humpback	30 000 +	ca. 3 000	10%	1, 2, 3, 12	—
Right	?	900- 3 000	?	1, 2, 12	±54%
Sperm: Male	ca. 257 000	ca. 128 000	50%	6, 12	± 6%
*** Female	ca. 330 000	259 000-295 000	78%		

Table 16 – Whale Populations in the North Pacific

Species	Estimated Original Population	Present Population Estimates	% of Original Population surviving (worst estimate)*	Sources	% Variation from mean of present Estimates
Fin	42 000- 45 000	14 000- 19 000	31%	7, 12	± 15%
Sei	58 000- 62 000	34 000- 38 000	58%	7, 12	± 5.5%
Blue	4 700- 5 000	1 400- 1 900	28%	8, 9, 12	± 15%
Humpback	?	1 200- 4 000	?	8, 12	±54%
Right	?	100- 1 000	?	8, 12	±82%
Grey	?	7 500- 13 000	?	10, 12	±27%
Sperm: Male	167 000-195 000	ca.69 000- 91 000	35%	11, 12	± 6%
*** Female	124 000-152 000	ca. 102 000 - 125 000	66%		

*   Worst estimate is obtained by comparing maximum possible original population with minimum possible present population.

**   This is probably far too pessimistic. The fact is that virtually nothing is known about Minke populations.

***   As pointed out on page 84 biomass should be examined also, particularly for Sperm Whales.

105

APPENDIX V

THE PROTECTED SPECIES OF WHALES

It is assumed that because some species of whales are totally protected by the IWC, that they will eventually recover to their original abundance. However, it must be remembered that this concept of reversibility is only an assumption, and we can only **hope** that the protected whales will recover. Even if we do accept the concept as a valid one, as whales are long lived, slow breeding species, their recovery period will be a long one. The Right Whales which were protected in 1935 have still shown little sign of recovery. This might be in part because of the continued aboriginal whaling. The scientific catch may also be a contributing factor as more than their present sustainable yield of Humpback Whales are currently taken by scientists for research. Also, factors such as inter-dependence between species and the competition for food must be taken into consideration.

It is worth remembering that it will take a protected, over-exploited stock considerably longer to recover to the level and composition at which it could sustain a maximum yield in weight than it would in numbers.

a) **BLUE WHALE, SIBBALD'S RORQUAL or SULPHUR BOTTOM WHALE**
   (*Balaenoptera musculus*)

The Blue Whale is the largest animal that has ever inhabited the earth. It can grow to over 100 feet in length, and can weigh up to 150 tons.

With the development of modern whaling techniques the Blue Whales soon became the prime target of the pelagic whaling fleets and from 1910 until it was finally protected in 1966, it was slaughtered relentlessly. It has been estimated that the number of Blue Whales slaughtered during this period was 330 000, with 29 400 in 1931 alone. The Norwegian industry, followed by the UK, were the main countries involved in the massacre, which did not cease until the Blue Whale was on the verge of biological extinction. However, it must be noted that both the Russians and particularly the Japanese pushed up the catch of Blue Whales at the critical time when the stocks were falling drastically.

The Blue Whale has been 'commercially extinct' for over a decade, and "there is not sufficient data to permit any assessment" (FAO/ACMRR 1974) of the present populations, because since the commercial hunting of Blue Whales was prohibited, very little research has been carried out on this species. The scientists now believe that the population estimates used in the early 1960s were probably too low – by a factor × 2, but numbers were still disastrously low. Had the Blue Whales been managed rationally they could have produced an annual revenue of around $65 million to the Japanese industry (1971 values).

A point about which there is still some doubt is the position of the Pygmy Blue Whale. This is claimed to belong to the same species as the Blue Whale, and consequently enabled the Japanese, in the latter days of the Blue Whale fishery, to increase the quotas. It must be borne in mind that as far as setting the quotas is concerned, it is utterly irrelevant whether or not the Pygmy Blue Whale is a full species; even if it only represents a subspecies, this will also represent a discrete (separate) population which should have the quota set separately from the other populations.

Figure 8   U.K. Antarctic Whale Catch (in tons) 1946-63

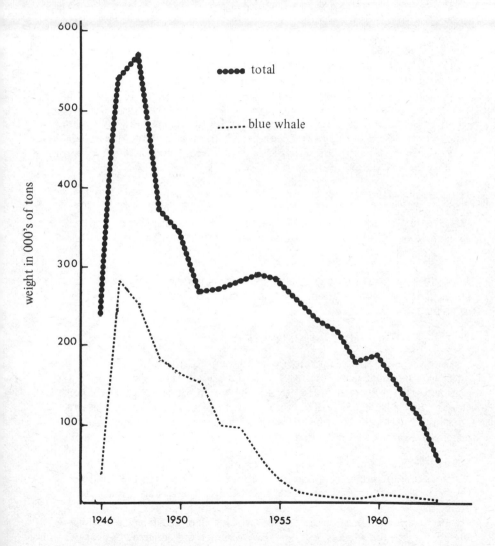

b)   **BLACK, BISCAYAN or SOUTHERN RIGHT WHALE**
     (*Balaena glacialis*)

The Black Right Whale (over 50 feet) is very similar in general appearance to the Greenland species, but with a smaller head, shorter baleen plates, and it has a distinctive, irregular prominence on the top of the head (the bonnet). Black Right Whales are found in temperate waters of the Pacific and Atlantic.

As far back as the 11th century it was hunted in the Bay of Biscay; the Basque whalers provided the foundations of modern whaling and eventually brought the species to near extinction in the North Atlantic.

At one time it was also abundant in the southern hemisphere; Sir James Clark Ross observed them in the Weddell Sea in 1844 and wrote: "We observed a very great number of the large-sized black whales, so tame that they allowed the ship sometimes to almost touch them before they would get out of the way; so that any number of ships might procure a cargo of oil in a short time. Thus . . . we have discovered . . . a valuable whale-fishery well worth the attention of our enterprising merchants!" But by the time a full-scale whaling expedition was sent out from Dundee in 1892, the Black Right Whales had already been depleted by the activities of British and American whalers hunting them on their migration routes.

There are no reliable estimates available at present for the North Atlantic populations. The North Pacific population has been estimated at ca. 250 (Wada, S., 1972, unpubl. rep. to IWC); the Southern hemisphere stocks, which may be increasing slowly are estimated at 4 300 (Masaki, Y., 1972, unpubl. rep. to IWC).

Recently they have been studied alive in South American coastal waters and Dr. Roger Payne's observations have shown that they indulge in 'sailing', using the tail flukes as a sail.

c)   **GREY WHALE** (*Eschrichtius robustus*)

The Grey Whale is the sole member of the family Eschrichtiidae and is the most primitive surviving baleen whale. Its structure is remarkably similar to extinct cetotheres, which were ancestral to all living baleen whales (Rice & Wolman[1]). Grey whales reach a length of about 45 feet. Like the Right Whale it lacks a dorsal fin, but does have some grooves on the throat. It travels the longest distance of any whale during its migrations — from 18 000 kilometres each way.

The Grey Whale formerly existed in two geographically isolated areas: in western North Pacific the population numbered about 1 000-1 500 in 1910 and was virtually exterminated between 1899 and 1933; in the Eastern Pacific the population originally numbered about 15 000, but during the second half of the 19th century, in particular, it was drastically depleted. The Grey Whale was an easy prey to the whalers as it is a slow swimmer and congregates near the shore. In 1947, when the population may have been as low as 250, it was given complete protection; the population is currently estimated at around 11 000 and has apparently stabilised at this figure. A small number (181 in 1972) are killed each year, mainly be aboriginal fisheries in Siberia and North America and some are taken by scientists (74 in 1969, California). In 1971 Rice and Wolman published a detailed account of the Grey Whale and found that during the years 1932-1960 the population had apparently increased at 11% per annum. In order for such an increase to occur they pointed out that it would be necessary to assume that virtually no adult mortality had occured, which is quite contrary to the theories of an average annual mortality postulated by Dr. R. Gambell and others.

Basic data published by Rice and Wolman includes: birth rate = 13%; annual mortality = .082%; sex ratio = 1:1 at all ages and 44% of the population is sexually mature.[1]

The Grey Whale is the object of the world's only whale sanctuary, created by the Mexican Government in 1972 in Scammon's Lagoon, Baja, California. It has become a popular tourist attraction, visitors travelling mainly from Los Angeles and San Francisco to see the whales on their breeding grounds, and the establishment of the reserve has regulated the activities of the cruise boats bringing tourists, though visitors are still thought to cause some disturbance.

d) GREENLAND RIGHT or BOWHEAD WHALE (*Balaena mysticetus*)

The Greenland Right Whale (up to 60 feet) is plump-bodied with an enormously large head. It lacks a dorsal fin and does not have the throat furrows of other rorquals. The head (nearly one third of the total length) has strongly arched jaws, long flexible baleen plates and a characteristic white area on the front of the lower jaw. It is found in Northern waters.

The Greenland Right Whale was one of the first whales to be hunted to commercial extinction, and was the centre of the Spitzbergen Industry which started in 1611. No commercial whaling has taken place since about 1915, but it continues to be hunted by eskimos on a fairly small scale.

In the past it was widely accepted that even in the event of a world-wide moratorium on commercial whaling, aboriginal fisheries such as those practised by the eskimos, should be allowed to continue. However, the eskimos can no longer claim to rely on whaling for their existence — many of them have even moved into a monetary economy. On the other hand, a cessation of whaling could adversely affect their culture. As Ed Mitchell's report states, "In some cases permitting aboriginal hunting of protected species no longer seems to be relevant to the survival of cultures. Some of these cultures (have) changed anyway and new equipment (is) being used to continue capturing whales".

The effect of the Canadian eskimo fishery on the Greenland Right Whale populations is unknown, but the inefficiency of their harpooning technique undoubtedly leads to a far greater mortality than the recorded kills would

---

(1) Source: Rice, Dale W. and Wolman, Allen A., 1971, The Life History & Ecology of the Grey Whale (Eschrichtius robustus), ASM Spec. Pub. no. 3 Washington.

suggest. It is likely that seriously depleted populations can be held down for many years by even a very small annual kill. In 1972 39 were killed in the US.

In 1977 the Scientific Committee of the IWC recommended a zero quota for this species. The Commission nonetheless set a quota.

e) HUMPBACK WHALE (*Megaptera novaeangliae*)

The Humpback Whale (up to 50 feet) is rather different in appearance from the other baleen whales. Like the rorquals it has a dorsal fin and grooves on the underside, but it lacks the streamlined appearance of the rorquals. It is very rounded, with extraordinarily long flippers; the flippers and head are covered with irregular callosities.

The Humpback was formerly an important part of the whaling industry, particularly to the shore stations operating during the early part of this century. No realistic estimates of the original population in the North Pacific are available, but in 1966 when it was given complete protection, it was thought that the population numbered about 1 200. Since that time no increase has been recorded. The North Atlantic population is less than 1 000, but may have increased slightly in recent years. The southern hemisphere populations which are thought to have once numbered somewhere in the region of 100 000 had been reduced to about 3 000 by the time they were protected; the populations do not show any sign of recovering.

APPENDIX VI

SMALL WHALES

It is often overlooked that a wide variety of small whales are taken by whalers, some commercial, some aboriginal, and, by far the largest number, as an incidental to the tuna and other fishing industries. In 1970-71 it has been estimated that about 388 000 Spotted Dolphins, 117 000 Spinner Dolphins (*Stenella* species) and 15 000 Common Dolphins (*Delphinus delphis*) were killed in the course of the USA tuna fishing operations. Efforts have been made under the US Marine Mammals Act 1972 to improve the methods of fishing and to cut the mortality. The other species killed are outlined below:

a) **COMMON PILOT WHALE or BLACKFISH** (*Globicephala melaena*)

From 1965-1970 Norwegian whalers killed 589; in the same period Canada killed 3 612 and Japan killed 1 172. In the Faroe fishery, a famous and long-established one, the Pilot Whales are herded in shore for a mass slaughter known as Ca'ing; from 1969-1973 3,326 were killed. Greenland and USSR also killed very small numbers. The Faroe records go back many years and it is estimated that from 1584 to 1883 some 177 000 were killed.

b) **NORTH PACIFIC BOTTLE-NOSED WHALE** (*Berardius bairdi*)

This species is normally only killed by Japanese whalers; from 1965-1971 936 were killed; and approximately 1000 were killed in February 1978.

c) **WHITE WHALE or BELUGA** (*Delphinapterus leucas*)

This Arctic species is still killed in some numbers, often by aboriginals. From 1965-1972, 2 902 were killed by Canadians. In the same period 6 757 were killed in Greenland waters. The only other area where significant numbers are killed appears to be the USA, where in 1972 the annual kill was estimated at 185.

d) **KILLER WHALE** (*Orcinus orca*)

The largest predatory species. Between 1965 and 1972 Norway killed 973 and Japan 470. Small numbers were also killed in Greenland, USA, Canada, the Faroes, Peru and in the Antarctic by USSR. In 1972 South Africa (Natal) killed 17.

e) **NORTHERN ATLANTIC BOTTLE-NOSED WHALE**
   (*Hyperoodon ampullatus*)

Norway is the only nation which appears to have taken significant numbers of this species in recent years, taking 1 250 between 1969 and 1972. USSR, Greenland and Canada have also killed small numbers. There is some indication that the species is declining. In the 1890s some 2 000-3 000 were taken each season.

f) BOTTLE-NOSED DOLPHIN (*Tursiops truncatus*)

Several hundred a year are taken off Florida and other parts of coastal USA for display in dolphinaria and for research. Small numbers are killed in the Lesser Antilles area by native hunters. Some are killed in the course of the tuna fishing industry in the Pacific.

This is the species most commonly displayed in public dolphinaria and is widely used for scientific and military research.

g) ATLANTIC WHITE-SIDED DOLPHIN (*Lagenorhynchus acutus*)

"This dolphin is taken only in summer in the waters of Norway . . . The take may reach 1 500 individuals at a time." Tomilin, A.G., 1957, Mammals of USSR Vol. IX.

h) SARAWAK DOLPHIN (*Lagenodelphis hosei*)

This species was first made known to science as recently as 1956 and was only known from a single skeleton collected in Borneo, until 25 were collected from a tuna purse seine net. Since then, several more have been killed in seine nets. Although the numbers are fairly small, because of its overall scarcity this mortality could affect the survival of the Sarawak Dolphin.

i) COMMON DOLPHIN (*Delphinus delphis*)

This is probably the most widespread of the world's dolphins. The Common Dolphin has been commercially hunted in many parts of the world, often only in small numbers. In 1940 it was estimated that 110 000-120 000 were killed each year in the Black Sea alone. The dolphin is protected by the Russians and Bulgarians, but not by the Turks, who still have a dolphin fishery. This fishery is subsidised by the Government and the oil from the dolphin is used for lighting. They are still hunted by some countries in the Mediterranean, the Azores and many other parts of the world. Large numbers are killed incidentally in the course of tuna fishing (15 000 in 1970-71).

j) NARWHAL (*Monodon monoceros*)

Although Canadians took 341 in 1973, the only nation regularly hunting this species appears to be Denmark (Greenland), which took 989 between 1969 and 1972.

k) COMMON or HARBOUR PORPOISE (*Phocoena phocoena*)

Between 1969 and 1972 Greenland is recorded as having taken 4 889 though others may well have been killed elsewhere. In 1957 it was reported that 2 000-2 500 were killed in single catches on the migration route between the Black Sea and the Sea of Azov.

Miscellaneous

Unidentified small whales have been killed in many places, though usually in small numbers. The greatest number seem to be killed by Greenland and Japan, the latter killing 759 from 1965 to 1971.

MAJOR SMALL WHALE FISHERIES

Some of the numbers of the smaller whales reported in the International Whaling Statistics are tabulated below:

Table 17 – Japan

Year	Pilot	Bairds	Killer	Misc.	Total
1965	288	172	169	71	700
1966	199	171	137	220	727
1967	237	107	101	294	739
1968	166	117	22	274	579
1969	130	138	16	56	340
1970	152	113	12	29	306
1971	181	118	10	25	334
1972	91	86	3	17	197
Total	1 444	1 022	470	986	3 922

Table 18 – Japan

Year	Pilot	White	Killer	Bottle-Nose	Misc.	Narwhal	Common Porpoise	Total
1965	194	481	1	—	1 365			2 041
1966	224	540	—	—	899	see below		1 663
1967	228	584	—	—	1 295			2 107
1968	—	1 210	—	—	1 970			3 180
1969	—	948	1	—	—	220	1 328	2 497
1970	—	1 416	1	1	2	417	1 184	3 021
1971	—	737	2	—	1	232	1 319	2 291
1972	1	841	—	—	—	120	1 047	1 909
Total	647	6 757	5	1	5 532	989	4 878	18 709

(For 1965-1968 the miscellaneous totals include Narwhal and Common Porpoise. The ratio of porpoise to Narwhal in later years is approximately 4:1. On that basis from 1965-68 approximately 4 426 Common Porpoises and 1 106 Narwhal were taken. Between 1968 and 1972 Greenland also took 1 070 Minke Whales.)

SMALL CETACEANS

In addition to those described a number of other small cetaceans are hunted in various parts of the world. Ed Mitchell divided the small whales into four groups, based on the needs of the various species for population assessment/degree of exploitation. Of the species not dealt with elsewhere in this manual, the following were grouped as being heavily fished and in urgent need of accurate population assessment:

Dall's Porpoise	(*Phocoenoides dalli*)
Spotted Dolphins	(*Stenella species*)

The following species are taken locally in small numbers, but this level of exploitation may still have a significant effect on the populations:

Cuvier's Beaked Whale	*Ziphius cavirostris*
Broad-beaked Dolphin	*Peponocephala electra*
Pigmy Killer Whale	*Feresa attenuata*
Short-finned Pilot Whale	*Globicephala macrorhyncha*
Atlantic White-sided Dolphin	*Lagenorhynchus acutus*
Gulf of California Porpoise	*Phocoena sinus*
Black Porpoise	*Phocoena spinipinnis*
Black Finless Porpoise	*Neophocaena phocaenoides*
White Flag Dolphin	*Lipotes vexillifer*
La Plata Dolphin	*Pontoporia blainvillei*

APPENDIX VII

WHALE PRODUCTS

As Table 19 shows, whale products are found in the most surprising places. Historically whales have been hunted primarily for their oil and, more often than not, the rest of the whale was simply thrown away. At the turn of this century, however, considerable efforts were made to create markets for the rest of the whale. The scope of table 19 indicates the success of these efforts.

Very few whale derivatives are currently used in Great Britain. In the FOE Whale Campaign Manual No. 1, we listed a number of firms ranging from cosmetics and pencil manufacturers to pet food and pharmaceutical manufacturers who used whale derivatives in their products. We are pleased to report that now there is a very slim chance of finding a product with whale derivatives in it on the shelves of our supermarkets or village stores.

This is for several reasons:

1) The cosmetics firms are phasing out the use of animal products because the source is unreliable and because better synthetics have been developed.

2) Whale meat is no longer used in pet foods manufactured in the UK because its import has been banned by the Government.

3) Many firms were genuinely unaware that the whale was endangered and that they were contributing to the whales' demise.

4) Adverse publicity.

5) The increasing cost of whale products.

Sperm oil is still used in industry, however, as a lubricant and cutting agent and in the textile and leather industries; but we believe that substitutes are available for most of these processes (see next section).

In other parts of Europe, whale oil is used in margarines and cooking fats, and it is likely that whale meat is used in pet foods. Spermaceti is used in cosmetics such as lipsticks and face creams, and ambergris in the most expensive scents. If the markets for these products were closed in Europe it would go a long way to closing down the whaling industry. The money from the sale of sperm oil by most whaling nations undoubtedly keeps them operating at a profit. Even though the Japanese say they need whale meat, we doubt whether they could continue to whale if their whale meat was not subsidised by the profit from the sale of sperm oil.

Diagram 3

PARTS OF THE WHALE USED COMMERCIALLY

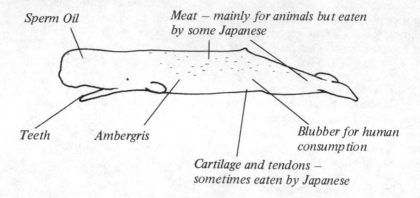

Sperm Oil

Meat – mainly for animals but eaten by some Japanese

Teeth *Ambergris*

Blubber for human consumption

Cartilage and tendons – sometimes eaten by Japanese

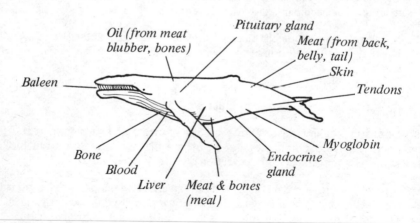

Oil (from meat blubber, bones)

Pituitary gland

Meat (from back, belly, tail)

Skin

Baleen

Tendons

Bone

Blood

Liver

Meat & bones (meal)

Endocrine gland

Myoglobin

Table 19 – Past and Present Commercial Uses of the Whale

(Whale by-products thought still to be used in the United Kingdom are marked with an *)

| Ambergris (from intestine of the Sperm Whale) | Fixative for scent* |
| | Used in high quality scented soap* |

Baleen	Bones for corsets, bustles and collars Whips and riding crops Umbrellas Hooped skirts Brooms Brushes
Blood	Added to adhesive in plywood manufacturers Fertilisers
Chemical salts	Creatine used in soups

Collagens (present in bone,
   skin and tendons which are
   boiled to yield gelatine)
	Gelatine is used for:
	   Photographic film
	   Edible jellies
	   Confectionery

Endocrine glands (yield hormones)	Medicines and pharmaceutics*
Liver	Whale oil from liver yields Vitamin A
Pituitary glands	ACTH (cortisone derivative) used in treatment of rheumatoid arthritis

Skin (from Toothed and
   White Whales)
	Leather for:
	   Bicycle saddles
	   Handbags and shoes*

Sperm oil:
   Unrefined sperm oil
	Mixed with mineral and other oils for
	   dressing hides in leather industry*

Refined and filtered (Spermaceti)	Cold creams* Lipsticks* Brushless shaving creams Ointments*
Filtered sperm oil	Ingredient of lubricating oil for light machinery*
Sulphurised sperm oil	Emulsifying agent in compounded oils* Cutting oils* Textile lubricants* Dressing hides*

Saponified sperm oil
   (sperm oil alcohols)
	Cetyl alcohol used as superfatting agent
	   in creams; also used on water-holes
	   in Australia to prevent evaporation.
	Oleyl alcohols:
	   Hair oils
	   Creams*
	   Lotions

Dye-solvents
Lipstick lubricants*
Free alcohols:
    Textile finishing*
    Dressing light leathers*
    Dye-solubilising and blending agents in printing inks
    Plasticiser bases for carbon papers and stencils
Oleyl and cetyl alcohols converted to Sodium salts of their sulphate esters:
    Powders
    Pastes
    Detergents
Cetyl alcohol converted to cetyl pyridinium bromide for cationic surface-active agents and germicides
Synthetic esters of cetyl and oleyl alcohols produced for plasticising, synethtic resins and emulsification work

Tendons	Tennis racket strings Surgical stitches
Whale bone	Bone meal for fertilisers Shoe horns Chess sets Toys
Whale meat	Ingredient in pet foods Frozen whale meat used for feed in mink fur farms Food for zoo animals Culture medium for screw-worm flies Soups and gravies Powdered whalemeat used for animal feedstuffs

Whale oil
(Glyceridic oil of the Baleen Whales is used primarily in the production of glycerine, margarines and soaps)
	Saponified — yields glycerine for:
	  Dynamite
	  Curing cigarette tobacco
	  Medicines
	— yields stearates for soap
	Polymerised — yields:
	  Varnishes
	  Oil cloth and linoleum

Drying oil in the manufacture of paints
Printing ink
Hydrogenated — yields:
Margarine
Cooking fat compounds
Lard
Shortening
Hardened Fin whale oil is used in:
Candles
Crayons*

Whale teeth (Toothed Whales such as Sperm)	Ivory for carvings and souvenirs Piano keys

In the UK the import of most whale products is not controlled, and this includes skin, bones and all made-up products. Where no controls exist there is no record of imports, so that the amount of trade in these goods is unknown.

Uses of whale products that have recently come to light in the UK include baleen in some archery gloves and, apparently, in the decoration of policemen's helmets. Also sperm oil is contained in the lubricant used by some bakeries on the blades which cut dough into loaf sizes.

APPENDIX VIII

SUBSTITUTES FOR SPERM OIL AND ITS DERIVATIVES

a)   SUBSTITUTES FOR SULPHURISED SPERM OIL LUBRICANTS

Sulphurised sperm oil has long been an important ingredient in metalworking and cutting oils because of its good emulsifiability and excellent lubricating properties under stress. It has also been an important additive to oils and greases for use at high temperatures and pressures.

The importance of these uses prompted many manufacturers in the US to seek alternatives to sperm oil derived lubricants when the US government announced its intention to ban the importation of whale products into that country. As a result of this research, a number of substitutes for sulphurised sperm oil have been found.*

Calber Chemical Inc., of Philadelphia, market a range of substitutes for sulphurised sperm oil, some of which are claimed to be superior to the natural product. They are all synthetic. One of their products, "CEPAD-1051", is specifically designed as a replacement for 9½-10½% sulphurised sperm oil in gear oil formulations and they claim that in most use-areas they have successfully changed over to synthetic materials. Although they do not have agents in this country, Calber service accounts from their Philadelphia headquarters (see address list p. 124).

Another American firm, the Mayco Oil and Chemical Co., markets a range of sulphurised sperm oil replacements containing varying amounts of sulphur. Compounds in this, the "Maysperm 2000" series, have been tested as ingredients in extreme pressure greases, gear oils and metalworking oils. The wear and extreme pressure characteristics of these additives are said to be equivalent to those of the corresponding sperm oil derivatives and their heat stabilities, corrosion resistances, etc. are all satisfactory. The "Maysperm 2000" series is available in Europe from Rademaker Olie Chemie of the Netherlands.

In Britain, Esso Chemicals (Paramins) have developed a synthetic replacement for sulphurised sperm oil for use as an extreme pressure additive in metalworking and cutting oils. This product, "Parapoid 40", is technically equivalent to sulphurised sperm oil in all the applications studied by Esso and in some respects — e.g. its metal corrosion and emulsifiability characteristics — it is superior. Although "Parapoid 40" is, at present, more expensive than sulphurised sperm oil, it seems likely that the price of sperm oil may soon overtake that of this substitute.

The Kell Chemical Co. US, have also developed a number of substitutes for sulphurised and chlorinated sperm oils; some of these substitutes are already

---

The latest information (1978) is that the following companies no longer use any sperm oil: Duckhams, BP, Mobil, Amoco, Murco and Gulf.

being used successfully in industry as extreme pressure additives. Their price-performance relationships are said to be equivalent to those of the naturally derived products but whether this would apply to these products if imported into the UK remains to be seen.

A product developed by Chevron Oil, namely OLOA 910, showed promise as a sperm oil replacement but is currently unavailable owing to a raw materials shortage. Fortunately, this shortage does not appear to have affected the availability of "Parapoid 40", "Maysperm 2000" or the Calber products.

As far back as 1972 we were informed by a leading lubricant manufacturer that alternatives to whale-derived metal working, cutting and quenching oils were available but that tradition had been the main stumbling block to their employment. Happily this obstacle seems to have disappeared, and as of 1975 we are able to say that motor car lubricants are now highly unlikely to contain sperm oil. Substitutes for oils for tractors gearboxes have been developed (Centlube Super Tractor Oil and Dentax W), for gears (Omala range, Parapoid 40, Maysperm 2000 range, Oloa 910 (in the USA only), Kell Chemical range, Cepad range); Cutting and Grinding oil substitutes (Duckhams have entirely phased out sperm whale oil in this application and two other replacements exist, Maysperm 2000 range and Parapoid 40); and Quenching oils. (This use has been almost entirely discontinued. Wilkinson Sword have recently abandoned using sperm oil for quenching.) We feel certain that a total change-over could be effected quite easily within 12 months were a total ban introduced.

b) SUBSTITUTES IN THE COSMETIC AND PHARMACEUTICAL INDUSTRIES

Sperm oil, the alcohols and esters derived from it, and spermaceti were all used extensively in the preparation of cosmetics, the most widely used whale product being spermaceti. Since the start of our campaign, the use of these products by the major manufacturers has stopped – for 'conservation' reasons and also because the quality and availability of the synthetic substitutes is more easily controllable than that of the natural products.

The substitutes which have been introduced range from vegetable waxes to synthetic esters. In some cases – e.g. that of cetyl alcohol – the product is the same but it is obtained from non-whale sources such as plants. The Toilet Preparations Federation lists the following materials as substitutes for whale products (excluding ambergris): vegetable derived cetyl alcohol; mineral and vegetable waxes; microcrystalline waxes; synthetic Sperm wax; synthetic straight-chain primary fatty acids (C_4–C_{20}); cetyl palmitate; tallow/vegetable based materials; and partial formulations of the materials from ingredients of non-whale origin.

The cosmetic industry in the United Kingdom has now all but totally abandoned the use of sperm oil and it is unlikely that even residual use continues. Cosmetics and face creams imported from abroad may well contain sperm oil. With the expected introduction in France of a total import ban on all Sperm Whale products, this picture may soon change. (Ambergris is still used by some British scent manufacturers, despite the existence of substitutes.)

c)   SPERM OIL IN THE TEXTILE INDUSTRY

The main use for Sperm oil in the textile industry has been as an ingredient of combing oils for worsted processing. In recent years, however, the use of combing oils has decreased and the Wool Industries Research Association has widened its combing oil certification scheme to allow for the introduction of alternative types of oil. The following types of oils are now considered for acceptance under this scheme. Sperm/castor oils; arachis oil; blended oils (with or without mineral oils); synthetic, polyoxyalkylene derivatives; and self-emulsifying mineral oils. Sperm/castor blends were the most popular a few years ago but these are apparently being replaced by the polyoxyalkylene oils. One such oil is "Oxitex 70", a Shell product marketed by Stephenson Bros. Ltd.

d)   SPERM OIL IN THE LEATHER INDUSTRY

Sperm oil's unique characteristics make it an ideal material for use in the leather processing and finishing industry. Sperm oil is "used in making a fat liquor for the treatment of leather where it has shown to be more or less unique in giving the handle to leather unaffected, relatively speaking, by extremes of temperature and weather". (Lankro Chemicals Ltd., 10 May 1972). In fact the Leather Industry, which may consume no more than 1,000 tonnes per annum of sperm oil, is the main objector to a prohibition on imports. Of all the industries it appears entrenched in the attitude that there can be no substitutes for sperm oil, yet many substitutes have been developed:*

Soft leathers, for gloving or for soft leather clothes:
Synthol CTA	*Bursoline SCO*
Lipoderm Liquor SAT	*Bursoline NAM*
Synthol LTA (more suitable for less stretchy uses)	
Tripon Products including BXE, DX, SNS, KX	
Sulfimex 214-5	

Other leathers:
Synthol SCT	*Synthol LTA*
Sulpho-Sternlic NS	*Sulpho-Sternlic MB*
Osipol range	*Tripon range*
Higate MC.11	*Higate G.150*
Lipoderm Liquor SAF	*Neatsofort oil*
Bursoline SCO, SM, NAM, SNCA	
Lipoderm Liquor SXA (industrial leathers)	

*Some are not used, on grounds of cost. The price of sulphited sperm oil rose 72% between February 1976 and February 1977, and the September 1977 price was £680 per tonne.

(Chamois leathers mostly do not contain sperm oil.)

Substitute research is bedevilled by the difficulty of finding a material to suit everyone's needs and methods of operating. Certain tanners have clearly devoted much time and money to adapting to substitutes; others — and it is their voices we hear now — appear to sit back and wait for someone to come up with the perfect alternative. It is worth noting that British-made gloves available from British Home Stores and Littlewoods are made without sperm oil, and that while the leather industry still claims that no substitutes are available for making high quality gloving leather, at least one of the gloving substitutes listed was considered suitable for very soft quality gloves, but too expensive.

The leather industry and the Government alike cannot produce useful figures for leather exports to justify continuing the import of sperm oil. The Foreign Office in 1975 went so far as to say that "The Sperm Whale which you mention is now, along with all the other species of Great Whale, in the category of a protected stock, which means that the taking of Sperm whales is not now permitted". (The 1975 Sperm whale quota was 19,040!)

e)   SPERMACETI IN PENCIL LEADS

Spermaceti has been used in pencil leads to make pencils smooth to write with. Berol Ltd, one of the leading manufacturers of pencils in this country, have informed us that they, have developed a product which is as a spermaceti substitute.

General

At the moment, there is no **single** product which can adequately replace sperm oil in all its applications, although suitable materials have been found to replace the natural product in all of its major use areas.

It would technically be possible to formulate sperm oil from its constituent chemicals, some of which could be extracted from plants and others prepared synthetically, but this would be extremely expensive. The most promising multi-purpose sperm oil substitute seems to be the oil extracted from the seeds of the slow-growing Jojoba shrub (*Simmondsia chinensis*). This plant grows in arid areas of North America and yields an oil which, like sperm oil, is really a liquid wax. Eventually, it will be produced by the American Indians, but it will be a long time before production can meet current use of sperm oil and, as we have suggested, by the time Jojoba is in large scale production sperm oil supplies may have dropped still further or ceased. A bibliography of papers dealing with Jojoba and related topics has been published by the Office of Arid Lands Studies, University of Arizona.

NAMES AND ADDRESSES

Berol Ltd., Northway House, High Road, London N20.
Calber Chemical Inc., East Ontario & Bath Streets, Philadelphis,
    Pennsylvania 19134, USA.
The Leather Institute, 9 Thomas Street, London SE1.
Lipo Chemical Inc., 114 East 32nd St., New York, NY 10016, USA.
Lubrizol Great Britain Ltd., Elliott House, Allington St., London W1.
Mayco Oil & Chemical Co., Beaver & Canal Sts., Bristol,
    Pennsylvania 19007, USA.
Office of Arid Lands Studies, University of Arizona, Tucson,
    Arizona 85719, USA.
Paramins, Esso Chemicals Ltd., Arundel Towers, Portland Terrace,
    Southampton SO9 2GW, Hants.
Rademaker Olie Chemie, Veenendaalkade 391, 2040 - Den Haag, Netherlands.
E. R. Squibb & Sons Ltd., Regal House, Twickenham TW1 3QT.
Stephenson Bros. Ltd., Listerhills Road, Bradford 7.
Tanners' Council of America Inc., 411 Fifth Ave., New York, NY 10016, USA.
The Toilet Preparations Federation Ltd., 22 Old Bond St., London W1X 3DA.
The Wool Industries Research Association, Headingley Lane, Leeds LS6 1BW.

Table 20 — Alternative Products

Products	Whale By-Products	Processing of By-Product	Possible Alternatives
Animal feeds	whale meat	meal of whale meat	Residual seed meal of Simmondsia. Meal of various wastes e.g. from sugarbeet, seaweed, meal, cereals
Candles - low quality	whale oil	hydrogenation	Beeswax
high quality	sperm oil & spermaceti	hydrogenation	Paraffin wax, Simmondsia wax, tallow
Cosmetics: lipsticks etc. cold creams	sperm oil & spermaceti	saponification	Essential oils such as lemon, orange, etc., Simmondsia oil, avocado cream, cactus cream, cucumber milk
Crayons & pencils	sperm oil, spermaceti & whale oil	hydrogenation	Simmondsia wax
Fertilisers	whale bones	grinding	Seaweed, various organic material, composted material

Product	Whale source	Process	Alternatives
Floor Coverings (linoleum & oilcloth)	sperm oil & whale oil	polymerisation	Linseed oil, Simmondsia oil
Glycerine	whale oil	saponification	Any saponified oil or fat, e.g. palm oil, ground nut oil.
Gelatine	skin, bones & tendons	boiling	Skin, bones, tendons & hooves of cattle, sheep, goats, etc.
Industrial oils:			
Cutting oil	sperm oil	hydrolisation & sulphuration	Linseed oil, castor bean oil, tung, Rapeseed oil, Simmondsia oil (and see p. 107)
Textile oil high speed	”	”	” ” ” ” ”
Machine oil	”	filtering	” ” ” ” ”
Watch & clock oil	”	”	” ” ” ” ”
Leather dressing	”	mixed with mineral oil	Various other available dressing oils (see p. 109)
Submarine oil	”	refining	Rapeseed oil, Simmondsia oil
High pressure gear grease	”	”	” ” ” ”
Automatic transmission fluid	”	”	” ” ” ”
Margarine	whale oil	hydrogenation	Vegetable oils such as ground nut oil, soya bean oil, sesame oil, maize oil, safflower oil, coconut oil.
Perfumes	ambergris		"Fixateur 404", other fixatives based on labadanum, oak moss, clary sage, cypress oils, agar, wood oil, etc.
Pet foods	whale meat	very little, if any	Fungal protein, abattoir waste, offal, cereal protein, soya bean meal
Pharmaceuticals:			
Ointments	spermaceti	refining	
Hormones	endocrine glands	and filtering	Domestic animal sources
Vitamin A	whale liver	oil extracted	Natural carotene from carrots

			& alfalfa; cod liver oil or synthesized Vit. A from lemon-grass oil or turpentine
Plywood glue	whale blood	dehydration	Traditional sources such as fish bone
Printing inks	sperm oil, whale oil	sulphurisation polymerisation	Simmondsia oil and Rapeseed oil
Shampoo	whale oil	cetyl alcohol derived from saponification	Fatty acid alcohols derived by saponification of other oils & fats, such as coconut or palm kernel oil
Soap	whale oil	saponification	numerous other oils such as palm oil.
Suntan oils	whale oil	cetyl alcohol derived from saponification	" " " "
Waxes for polishes and textile industry	sperm oil, whale oil		Simmondsia wax

APPENDIX IX

THE INTERNATIONAL TRADE IN WHALE PRODUCTS

There is a considerable trade in whale products in most countries of the world.
The only possible exceptions are New Zealand and the USA. The USA, banned
the import of whale products in 1971; however, they stock-piled sperm oil
before the ban came into force and has subsequently re-exported some of this
oil to Britain and other countries. Moreover, as some of the following tables
show, the USA was still importing whale products in 1974, despite the tightening
of legislation.

Table 21 shows the export value of whale products to the countries engaged in
whaling. (These figures do not include the value on the home markets.)

Table 21 — Export value of whale products: Oil (incl. Sperm)
(figures shown in thousands of US dollars)

	1969	1970	1971	1972
USSR	9 119	8 683	3 777	3 606
Japan	4 156	7 214	13 388	5 054
South Africa*	1 004	1 233	1 184	1 712
Canada*	374	218	587	395
USA*	465	707	720	106
Norway	554	648	804	285
Australia	562	550	1 361	1 208
Iceland	248	1 082	614	160
Portugal	325	220	400	154
Peru	198	293	819	314
Total	17 005	20 848	23 654	12 994

Source: FAO Yearbook of Fishery Statistics, Vol. 33, 1971.
*No longer engaged in commercial whaling.

Imports of Whale Products into Britain

Although the British Government has banned the import of some baleen whale
products and Sperm Whale meat, we still continue to import sperm oil, including
large quantities which originate from non-IWC countries in spite of voting in favour
of an IWC resolution, in June 1977, not to do this (see Table 22). Up to September
1974, we imported over 8,000 metric tons of baleen and Sperm Whale oil at a
value of over £1¼ million. Surprisingly, we were still importing baleen whale oil
in 1974 — 50.9 metric tons — despite the fact that it was banned by the
Government in March 1973. As can be seen from the table below, most of our
imports come from Australia and South Africa — 50.1% and 30.1% respectively.

Imports (by weight) of baleen and sperm oil products from the following countries during 1974 (Jan-Sept) were: 50% Australia, 30% South Africa, 12% Japan, 4% Norway and 3% Netherlands.

In Europe, one of the main importers (and re-exporters) of whale products is the Netherlands (which rejoined the IWC in 1977) and which also exports to the United Kingdom.

Table 22 (includes old Table 23) – **IMPORTS OF WHALE OILS INTO BRITAIN**
(Figures after 1974 reflect refined and unrefined sperm oil combined)

Origin	1973 quantity tonnes	1973 value £	1974 quantity tonnes	1974 value £	1975 quantity tonnes	1975 value £	1976φ quantity tonnes	1976φ value £
Iceland			700u	147,052				
Netherlands	1,002u	158,997	98.1r	28,758	1,016	196,373	500	113,825
			59.8u	15,538				
S. Africa	441u	54,782	2,202.2r	460,924	2,793.1	608,555	498	105,951
	2,020r	251,993	803.3u	102,115				
Japan	*7u	715	1,119.3u	226,939				
Australia	3,052u	306,976	6,027.2u	796,254	3,877.8	632,646	3,606.8	580,679
USA	1u	200						
	0.5r	135						
Norway	10r	1,222	347.6r	82,662	41.1	11,058	700	191,998
	22b	2,424						
W. Germany			98.1r	28,758				
France			94.1r	25,261	0.05	102		
Ireland	20b	2,800						
TOTAL	6,575.5	780,244	11,549.7	2,105,609	7,728.1	1,448,734	to August: 5,304.8	992,453
							whole year: 8,527.7	2,078,344

r Refined Sperm Oil
u Unrefined Sperm Oil
b Baleen Whale Oil
* According to Japanese trade figures, nothing was exported to the UK in 1973.
φ Figures for 1976 are incomplete as HM Customs and Excise refused to supply information on imports by country of origin after August 1976.

Source: HM Customs & Excise Statistical Office

Total sperm oil imports for 1977, according to figures supplied in February 1978
are: total quantity: 5,196.25 tonnes
total value: £2,477.204

Table 24 (includes old Table 25) – EXPORT OF REFINED AND UNREFINED SPERM WHALE OIL FROM BRITAIN
(Figures after 1974 reflect refined and unrefined sperm oil combined)

Destination	1973 quantity tonnes	1973 value £	1974 quantity tonnes	1974 value £	1975 quantity tonnes	1975 value £	1976 (to August) quantity tonnes	1976 (to August) value £
Argentina	12.4	2,210	24	3,778				
Australia	16.4	3,179	2	462	21.1	4,257		
Austria	147.1	21,445	171.3	32,873	56.2	15,066	1.8	680
Bahrain								
Belgium	91.3	16,050	230.5	53,922	141.2	44,602		
	37.1u	6,673	36.9u	10,756				
Brazil			39.0u	9,721				
Cameroon					0.7	204		
Canada	236.9	36,473	78.5	16,233	19.5	5,388	29.2	8,472
			10.2u	2,266				
China	28.6	2,455						
Colombia					90.3	21,543		
Congo					4.1	1,221		
Costa Rica	2.1	490					0.4	142
Cuba			4.3	1,895				
Curacao			3.1	621				

130

Country	1	2	3	4	5	6	7	8
Denmark	3.0	614	4.1	892	1.78	316		
Dominican Republic								
Dubai			19.9	3,927				
Finland	8.8	1,677	30.3	7,530	17.1	5,729	0.2	100
			3.0u	696			10.7	3,378
France	117.0	21,082	110.0	23,062	26.4	7,639	70.8	24,373
			19.6u	3,543				
Greece	4.0	726					2.0	690
Hong Kong							21.5	8.356
India	11.9	2,213	26.0	7,845	31.0	7,495		
			4.0u	1,476	24.8	8,086		
Iran	1.4	294	5.0	1,368	23.0	8,829		
Ireland			22.0	5,738	1.1	334		
Italy	490.2	89,308	421.2	101,938	399.3	110,391	258.6	69.870
			63.2u	14,370				
Kenya	18.3	3,679	118.1	29,175	52.5	16,911	7.2	1,210
Libya								
Luxembourg	26.3	2,700						
Malaysia	1.0	207						
Mexico	117.1	21,442	168.5	52,778	4.0	1,314	87.3	33,701
	10.3u	1,707						
Morocco	234.8	35,069	207.9	41,799	4.7	1,618		
Netherlands	18.9u	2,059	28.8u	5,984	566.6	145,170	851.3	227,577
New Zealand	10.6	2,134	5.3	1,071	2.2	661	12.6	5,410
Nigeria					1.0	580		
Norway	114.9		2.0	758	0.4	131	1,056	189,356
Pakistan	14.4	3,201	3.9	881	4.8	1,288		
Portugal								
Peru			1.4	392	2.3	680	0.2	337

Senegal	5.2				0.7			
Singapore	26.6	1,035			10.0	199		
Spain	24.2	5,331	68.2	15,628	4.9	2,117	1.1	150
Sweden		4,556	44.3	9,327	19.2	1,619	18.0	5,113
Sudan					0.8	5,985		
Switzerland	100.9	14,405	61.8 42.0u	14,371 10,443	17.8	277 5,033	29.9	8.426
Taiwan							1.0	288
Thailand	12.0	1,930	7.9	1,764				
Trinidad	6.6	1,020	4.5	1,111				
Turkey	5.2	1,020	12.0	2,582				
	4.0u	778						
Venezuela	14.9	2,905	9.6 5.6u	1,947 1,253				
W. Germany	478.0 49.1u	67,274 9,059	348.0	60,084	360.4	92,314	164.0	45,942
TOTAL	1,501.5	400,728	2,562.9	579,024	1,815.9	496,793	2,623.8	633,571

u Unrefined

Exports

Britain also exports large quantities of oils, mainly as refined oil. In 1974, total exports amounted to £428,406 as opposed to an import bill of £1,237,648.

The main importers of sperm oil from Britain appear to be Norway (taking 1,056 tons in 1976, January-August only), the Netherlands (851 tons, January-August 1976), Italy (484 tons, 1974), West Germany (527 tons, 1973, decreasing to 360 tons, 1975), with France and Mexico taking around 100 tons each per annum.

If Britain banned the import of sperm oil, the effects on the Australian whaling industry would be immediate. The importance of the British export market lies in the fact that Britain is one of the main refiners. Unless alternative outlets could be found the Australian industry would not be viable.

EEC Trade in Whale Products

The most recent trade figures were not always available for the EEC member countries. Tables 26 and 27 show the imports of whale products for 1971 and 1972. It can be seen that there is considerable circulatory internal trade which makes it difficult to assess the sources. The statistics for the individual countries do not always tally.

Table 26 — EEC Imports of Whale Products 1971 (weight metric tons)

Country of Export	Destination				
	France	Belgium/ Luxembourg	Netherlands	Germany	Italy
Netherlands	25	1 068	—	1 413	405
France		30	2		462
West Germany	66	1	751	—	24
Iceland	—	—	225	—	—
United Kingdom	148	78	60	171	167
Norway	1 150	31	25	3 012	1 488
Portugal	175	—	—	75	51
Spain	60	—	—	—	21
USSR	576	—	86	5 343	421
South Africa	81	—	535	—	10
Canada	—	—	—	747	—
Peru	—	—	—	1 005	299
Japan	303	—	7 633	13 613	40
Australia	575	—	—	101	—
Misc.	—	—	549	—	—
Total	3 159	1 208	9 866	25 480	3 388

Table 27 − Imports of Whale Products 1972 (weight metric tons)

Country of Export	Destination				
	France	Belgium/ Luxembourg	Netherlands	Germany	Italy
France	−	22	84	597	527
Netherlands	52	977	−	2 131	623
Germany	80	76	2 439	−	39
United Kingdom	166	6	323	392	371
Ireland	−	−	100	−	−
Norway	585	49	18	2 188	1 494
Yugoslavia	−	−	−	−	51
USSR	4 304	−	117	2 039	806
South Africa	404	−	5	785	−
Canada	−	−	336	951	−
Peru	−	−	−	459	400
Japan	596	−	437	14 208	−
Portugal	20	−	−	−	32
Turkey	300	−	−	−	−
Total	6 507	1 130	3 859	23 750	4 343

Table 28 − German (FDR) Imports of Sperm Oil Products 1973

Country of Export	Weight (metric tons)	Value (DM)
France	923	1 056 000
Belgium/Luxembourg	63	69 000
Netherlands	1 270	1 301 000
United Kingdom	619	627 000 (1)
Norway	1 254	1 293 000
USSR	1 680	1 515 000
South Africa	1 146	1 077 000
Japan	80	77 000 (2)
USA	175	108 000 (3)
Total	7 210	7 123 000

Source: German Trading Statistics 1974.

(1) According to British trade figures, only 49 tons were exported to Germany in 1973.
(2) According to Japanese export statistics, no sperm whale oil was exported to Germany in 1973.
(3) The export of whale products should be illegal in this instance

134

Table 29 – French Imports of Whale Oils 1973

Country of Origin	Weight (metric tons)	Value (French francs)
USA	1 083	2 128 000
Japan	166 (1)	340 000
Misc.	182	404 000
Total	1 431	2 872 000

Source: French trade statistics

(1)   According to Japanese export figures, no whale oil was exported to France in 1973.

Table 30 – Japanese Trade in Whale Derivatives 1973

Destination	Imports Weight (metric tons)	Value (000 yen)	Exports Weight (metric tons)	Value (000 yen)
Rep. Korea	733	229 209	1 227	128 066
North Korea	66	8 097		
China	10	956		
Iceland	2 920	605 209		
Spain	1 967	259 396		
USSR	16 539	1 464 320		
Canada	525	95 083		
Peru	2 573	50 027		
Central Africa	40	4 978		
Angola	583	129 146		
South Africa	504	61 063		
USA	4	1 198	1 018	53 666
Norway	27	7 228		
UK	750	17 471 (1)		
Netherlands			18 262	1 511 145
Taiwan			52	8 928
Philippines			5	853
Thailand			0.9	240
Total	26 566	2 933 381	20 565	1 702 898

Source: Japanese Trade Statistics 1974.

(1)   British trade statistics apparently do not include this figure.

Table 30 shows the Japanese trade in whale derivatives for 1973. It can be seen that the figures published by the various national trade organisations do not agree.

This is probably due, in most cases, to the fact that exporters often give the final destination of the products in their declarations, and the importers, the country from which it was last exported. Hence a cargo of oil sent from Japan to Holland via Germany will appear in Dutch statistics only as imported from Germany.

APPENDIX X

ALTERNATIVES TO WHALING IN THE ANTARCTIC

a)   KRILL

The Significance of Krill

Krill is not only significant in that it constitutes the basic diet of the baleen whales in the Antarctic, but also because it will undoubtedly be of supreme importance as a protein source in the future.

Burukovskii (1965) estimated that krill has a potential at least as great, in terms of weight, as all the presently exploited fish stocks combined. J. A. Gulland, in his book "Development of the Resources of the Antarctic Seas", also predicted that while it is desirable to achieve the proper management of whale stocks and procure from them their potential yield, the major future interest in the Antarctic must be in other animals.

Composition of Krill

Krill is the collective term for small shrimp-like planktonic crustacea. The kind which live in the Southern Ocean and which form the basis of the baleen whale's diet are *Euphausia superba.* They are about 6 cms. long and are found near the surface of the sea as they feed on the phytoplankton (floating plants) which live near the surface where they can get the most light. Krill are filter feeders and trap their food on the feathery growths at the base of their legs. The distribution of krill is uneven. They often gather in vast swarms — sometimes up to a quarter of a mile long and 200 yards wide, but it is thought that these high concentrations of krill do not cover wide areas, and that narrow bands of krill move southwards with the receding ice.

Euphausia superba is unusual among planktonic crustacea in taking more than one year to complete its cycle. Its life-span has been estimated to be 4-5 years and it reaches maturity and can first spawn at 2-2½ years. Its mortality rate during the first year is thought to be high, as carnivorous zooplankton take krill at this time.

In other oceans, krill consists of close relatives of *Euphausia - Thysanoessa* and *Meganyctiphanes* or other small crustaceans such as copepods and amphipods. In the sub-Antarctic, whales eat "lobster krill", a crustacean called Munida. The main food of the Greenland Right Whale in the Arctic is Clione, a sea-butterfly (mollusc) that feeds on other small animals. Whales also eat many other animals such as herring and squid. Penguins and cormorants have also been found in whales' stomachs.

What Animals eat Krill?

Besides the baleen whales, penguins, seabirds, seals, squid and shoals of fish also eat krill.

Quantities eaten by Whales

Blue Whales are estimated to eat 4 tons (about 8 million krill) and Fin Whales just under 3 tons of krill a day. Robert Burton (1973) estimated that before whales were exploited in the Antarctic, they consumed about 150 million tons of krill each summer. (Gulland estimated a lower figure of 50 million tons a year.) Now that the whale populations have been so reduced, Burton has calculated that the uneaten potential surplus of krill amounts to something like 100-150 million tons.

Harvesting the Surplus

This is not as easy as it might first appear. The processing of krill is even higher in energy consumption than is whaling and pelagic fishing. It consumes a great deal of non-renewable energy (i.e. oil) for the amount of food energy it produces. Once the krill has been collected the chitinous skin has to be separated. The remains then have to be converted into an edible form.

The uncontrolled exploitation of krill is indeed an alarming proposition for two main reasons:

1) because the importance of krill in the diet of whales means that such a fishery is likely to have a deleterious effect for them, especially since some species are already endangered; and

2) because over-exploitation could have a disastrous effect on the whole Antarctic marine ecosystem. Furthermore, as R. E. Longton pointed out, the experience of whaling has shown that once an industry with its huge capital investment has developed, it is too late to exercise effective control.

Other problems are connected with harvesting of krill as virtually nothing is known of the factors controlling its abundance or of the animals eating it (Gulland). Scientists do not yet know which sections of the population should be taken, and even if they did know, it would be extremely difficult to select in the field.

Potential Production

Gulland has calculated some rough estimates of the annual production of krill; these vary between 50 and 500 million tons. Lyubimova (1973) has also done some rather varied calculations and estimated that the Southern Ocean population amounts to 0.8 to 5 billion tons; and that the potential commercial catch is between 25 - 50 million tons.

Gulland has made the point that the sustainable yield cannot be calculated accurately until harvesting occurs at a level sufficiently high to have some detectable effect on the stocks. This seems rather like putting the cart before the horse.

The Countries harvesting Krill

The USSR has been harvesting krill for several years; having helped to virtually wipe out the baleen whales which have evolved over tens of millions of years, it would seem that they are now trying to take their place. Unlike baleen whales, ships do not have a function in the ocean life cycles and do not return nutrients, via excreta, to the oceans for other organisms to thrive on.

Krill paste and other krill products have been in commercial production in the USSR since 1970. We understand however that processed krill has not found a ready market there.

Soviet midwater trawls have produced harvests of 10-12 tons an hour, and processing has produced food protein paste and fodder meal at rates of up to 30% and 10% of the raw krill input respectively. The processed paste contains 70-75% protein, 13-20% fat, 3-10% carbohydrates and some 17 amino acids (Marine Pollution Bulletin 1973, vol. 4 no. 4).

Krill fishing expeditions have been mounted by the Japanese, Poles, Norwegians and West Germans, among others, and krill was the star of the show at an exhibition in 1977. Despite the enormous investment required, it seems very likely that krill fisheries will develop soon. The Greenpeace Japan Mission (1976-77) have suggested that the Japanese continue whaling so as to maintain their options for the krill fishery.

Management Problems

As J. W. Horwood pointed out in 1976[1], "the economic criteria of the (krill and whale) fishery have important consequences for its management". So it is that economic arguments, as so often in the past, may blur the view of other considerations of fishing krill. Little is known about the role of krill in maintaining the viability of the marine ecosystem, and it is therefore difficult to make rational judgement about its best use. If krill is to be harvested at all it should be jointly managed with the whales, for the populations of each heavily influence those of the other. The two fisheries, operating independently, must be in conflict.

This is not to say that the moratorium on whaling should not be the present priority. Whales are not the only krill eaters; indeed the depletion of baleen whale stocks has coincided with an enormous increase in populations of Antarctic fur seals, and crab-eater seals. Seals now eat more krill in the Antarctic than do baleen whales. It is therefore possible that seals may already be interfering with the populations of whales. If man takes krill, it is anyone's guess what will happen. For example, the amount of food available to slowly-recovering protected stocks might further slow their rate of recovery. In addition, as suggested by J. W. Horwood, the early maturity and increased pregnancy rate

---

(1) Paper "On the management of krill and whales" to the FAO Scientific Consultation on Marine Mammals, 1976.

in baleen whales may arise from an increase of food per animal; in this case much of the 'surplus' of krill is already being harvested by the whales.

Certainly a persuasive argument for the exploitation of krill can be put by any one country considering only its own short term interests. But in establishing its own right to exploit the resource, the right of other countries to do likewise is established. As with so many other "commons" the total of all the individual benefits does not necessarily add up to a benefit for the community as a whole, and especially over the long term. If the slaughter of the whales has taught us anything it should have taught us that this is true.

b)   SALMON

Most of the larger animals found in the Antarctic are dependent on krill, either directly, in the case of fish or indirectly, in the case of the penguins, seals, etc. which feed on the fish. Fish are already being harvested in the Antarctic but perhaps the most interesting development in recent years was the suggestion by Joyner, Mahnten and Clark [1] that salmon could be introduced as a means of converting the krill into an acceptable form of protein. The idea is that in selected rivers in South America salmon fry would be released and that, when they had undergone their migrations and growth period, they would return to their birth place, where they would be caught. This would have considerable advantages regarding energy compared with deep sea fishing. One of the main problems normally incumbent on any introduction of exotic species is the danger of their getting out of control, but with salmon, because of their migratory behaviour, and the fact that they are dependent on a limited habitat for breeding, it could be an easy matter to eradicate them should the experiment go wrong. Similar experiments have been carried out with transplanting salmon in the USSR and have met with a degree of success.

---

(1) Salmon, A Future Harvest from the Antarctic Ocean" by Timothy Joyner, Conrad V. W. Mahnken and Robert C. Clark Jnr., MFR paper 1063 Marine Fisheries Review, Vol. 36 No. 5.

RESEARCH ON CETACEANS

Most of the research in the past has been carried out on the bodies of whales taken in the course of commercial whaling. This is still true — even today. It is naturally desirable that while whaling continues scientists utilise every scrap of information they can glean from the slaughter of whales. However, it is also essential that research should take a new direction — towards the study of living whales.

Drs. Schevill and Allen point out in their paper, "Expanded Cetacean Research" (24th Report of the IWC), that this kind of research must be done so that the "place of whales in the ecosystem is better understood. We do not yet fully appreciate the likely consequences of disturbing the equilibrium by reducing or removing such important components as the stocks of whales. The information obtained from these studies would greatly broaden our understanding of the whale populations, their relationship to their environment and their responses to exploitation."

Many people have argued that if the 10 year moratorium on all commercial whaling was introduced, the scientific research that could be done on whales would be severely reduced. Schevill and Allen itemised the work which they felt could be usefully done during the period of the moratorium; some of their suggestions are listed below:

1)   Separate populations (stocks) of the different species of whales need to be recognised.

2)   The seasonal and geographic ranges of the different species and local populations need to be recognised.

3)   Marking techniques need to be improved; at present much crucial marking data from the national programmes is neither being submitted to the international 'clearing house' (BIWS) nor being published fully enough to permit scientific analysis (FAO/ACMRR Working Party on Marine Mammals Report, 1974).

4)   Expanded comparative studies of food and feeding need to be made to help our better understanding of competition between different species of whale.

5)   Study of individual whales; Dr. Roger Payne has found that the colour and pattern of individual whales remains relatively constant once the animal has reached maturity. The voice prints of individuals and populations require much closer study also.

6)   Small-scale scientific sampling cruises which would collect a small number of specimens for research and could carry out the following work:

   a)   biochemical studies of population genetics (using immunological and electrophoretic techniques, which have already been tried with some success).

b)   the study of contaminants of whale tissue — DDT etc. — as an index of marine pollution.

c)   chromosone studies, which would help our understanding of the evolution and systematics of whales.

d)   marking by means of persistent dyes which would enable scientists to establish the relationship between age and the structures used in age determination (such as ear plugs and baleen plates).

Studies of this type would help us to understand the social structure of whale populations, their behaviour, life history, abundance and their relationship with the rest of the marine environment. It would also have an advantage over the data supplied by commercial whalers in that we would be able to obtain information about whales in other seasons and other parts of the world. Thus far from impoverishing research, the quality and value could be greatly improved. However, there are two problems: first, research of any type requires money, and if there is no whaling industry to support the research, who is to foot the bill?

As whales are a "common heritage resource" it seems only appropriate that all countries should be prepared to fund the research needed to ensure its continued availability. However, as the IWC countries have in the past claimed special rights with regard to whales it seems only just that they should undertake special responsibilities in the present. We would like to see an international fund established to finance cetacean research and we feel that the major contributions to this fund should come from the IWC nations. At present the British contribution to research is pitiful, although in the past Britain made some contribution towards research: in 1923 the revenue of licences etc. in the Antarctic produced a surplus of £315,795 over expenditure and, of this, £300,302 was set aside for research.

In 1925 the "Discovery" set out from Portsmouth largely financed in this way. Such a fund could be administered by the United Nations and its existing agencies. Perhaps the most promising research being carried out at present is that of the FAO/ACMRR Working Party, which is critically reviewing the available data. The United Nations Environmental Programme has contributed some funds towards this.

The second problem is the lack of competent scientists to carry out the research. This problem would undoubtedly be solved rapidly were sufficient funds available.

There is one form of research that would require very little money and that is the evaluation and interpretation of the available data and samples now preserved in museums and institutions all over the world.

Current Research Projects

Although most research on live cetaceans has been carried out on the smaller species in captivity, the research which has so far been carried out on the great whales shows that continued commercial whaling is not essential for further research into whales.

Canadian researchers have been developing modifications of whale marks. The numbered whale marks, originally developed by the "Discovery" scientists in Great Britain have been widely used and have been useful in providing information on migrations and range of whales. The modified mark now being used by Canadian scientists has durable plastic streamers attached to it, which are visible at distances of up to a mile. The Canadians have also experimented with including small amounts of a harmless stain in the marks, which enters the blood stream of the animal and it is hoped will stain the ear plug and baleen plates, so that, should the whale be subsequently killed, a positive check on the rate of deposition of the laminations can be made. The stain used is the drug quinacrine which is fluorescent and detectable in concentrations of 1:5 000 000. In 1969 the Canadians also experimented with radio tags but, due to technical difficulties, the design was never perfected.

American scientists have tried using a balloon as an observation platform for observing Grey Whales. Although it was apparently a good method, blimps (at $700 an hour) were not easily available. In 1973 the National Aeronautics and Space Administration (NASA) cooperated in experiments with aerial remote sensing systems for detecting Grey Whales, but the high cost and only limited success have ruled this method out of general use.

The Norwegians appear to carry out little, if any, research outside the industry.

The South Africans have carried out research into Sperm Whales (British scientists have also been involved), but all this research was, at least indirectly, involved with the whaling industry.

Japan carries out some interesting research, but again mainly on dead whales. They have studied blood groups and various biochemical differences in the whale populations.

Information on research in the USSR is lacking; the little that has been published is mainly concerned with dead whales.

Dr. Roger Payne has done extensive research into the voice patterns of whales, especially Humpback Whales, and has made a record of their songs. He has also studied the Southern Right Whales off Patagonia.

APPENDIX XII

THE WHALE CONSERVATION MOVEMENT

Concern over the future of whales is nothing new. Scientists and whalers have made it clear on numerous occasions that they thought that stricter conservation measures were needed, but almost without exception they have gone unheeded.

In 1931, a paper published in the proceedings of the Linnaean Society of London by the world's foremost whale scientists of the time, Sir Sidney Harmer, started as follows:

> "In the Address which I delivered to this Society in 1928 I expressed my anxiety with regard to the effect of whaling operations on the stock of whales; and I alluded to the increased use of pelagic whaling, which seemed likely to result in a depletion of the stock within a short period of years. Much has happened since 1928; and, if anxiety was then justifiable, the events of the last two years may fairly be described as alarming."

E. Keble Chatterton, in Whalers and Whaling, written while the last whaling ship under sail survived, wrote in 1930:

> "And alongside our optimism in regard to the Antarctic we must place the plain observed fact that when a species of whale has by over-fishing been wiped out of any area, he does not go back to that region even long years after the fleets have given up using those waters. Of this we may quote two historical instances. For centuries the whale was hunted in the Bay of Biscay, right up till the sixteenth century, when it was about to be exterminated, had not the Greenland whaling grounds been discovered. To this day the Biscayan whale has never recovered its former numerical preponderance. Similarly the Finmark rorqual was hunted to extermination so that the industry came to a full stop after a few intensive years. If, then, the whales should be exterminated in the waters of the Falkland Dependencies, it is improbable that even after a respite they would return . . . indeed, there are those who believe that by the time the twentieth century has run out, there will be no more whales to hunt: they will become as past and banished as the prehistoric animals are to us today."

Dr. N. A. Mackintosh, a former head of the National Institute of Oceanography and for a long time the British Government's senior adviser on whaling, who was on the IWC for many years, wrote in 1965:

> "Great efforts have been made to reach agreement on measures adequate to conserve the southern stocks, but for some years it has been evident that the restrictions put on the catches were not enough to prevent progressive reduction of the stocks. Now (in 1964) the situation has sharply deteriorated, for at its latest meeting the Commission was unable to secure agreement to lower the catch limit to a level even

approaching the maximum required by the scientific evidence to halt a further decline of the stocks. The only species now receiving adequate protection are those which have already been reduced to such small numbers that they are commercially negligible. The outlook may have changed for better or for worse by the time this book is in print, but future developments are at present obscure." (The Stocks of Whales)

Dr. Mackintosh's pessimism was well founded. In 1973 Robert Burton in The Life and Death of Whales carried on the story:

"The western world, at least, is becoming conservation-minded but it is still difficult to show concrete examples of a radical change of attitude towards conservation problems by the people with the power to act, even when these problems concern our daily lives. The fate of whales is far from home; until recently it received little publicity and made no impact, so it has not been on the public conscience. Then in June 1972, the UN Conference on the Human Environment recommended a ten-year ban on whaling. This received considerable publicity but, at the International Whaling Commission meeting shortly afterwards the ban was rejected. On the other hand, whaling had now become the subject of public discussion."

Pressure groups have certainly had an effect in persuading Governments to take action on whaling. For example, American 'Save the Whale' groups were instrumental in getting an import ban on eight species of whales in 1971. Their success has precipitated other groups all over the world to take similar action. There is not enough space to mention all the work of all the individual groups, but some names and addresses are listed below:

Whale Conservation Groups

Friends of the Earth organisations:

Australia	*– FoE Sydney, c/o NSW Environment Centre, 263B The Broadway, Broadway, NSW 2007*
France	*– Remi Parmentier, 117 avenue de Choisy, 75013, Paris*
New Zealand	*– Philip Alpers, PO Box 39065, Auckland West*
Sweden	*– Jordens Vanner, Box 11107, S-10061, Stockholm 11*
United Kingdom	*– 9 Poland Street, W1V 3DG (01-434 1684)*
USA	*– Anne Wickham, 620C Street, SE, Washington DC 20003*

British Groups:

Fauna Preservation Society, c/o Zoological Gardens, Regent's Park, London NW1

(01-586 0872).

Greenpeace, 47 Whitehall, London SW1 (01-839 2093).

International Society for the Protection of Animals, Nick Carter, 106 Jermyn Street, London SW1 (telephone 01-839 3066).

Royal Society for the Prevention of Cruelty to Animals, The Manor House, The Causeway, Horsham, Sussex. (Horsham 64181).

World Wildlife Fund, 29 Greville Street, London EC1N 8AX. (01-404 5691).

Some Organisations working outside Great Britain:

Allied Whale, c/o College of the Atlantic, Eden Street, Bar Harbor, Maine 04609, USA.

ACT, Dr. James Mead, c/o Marine Mammal Study Centre, Smithsonian Institute, Washington DC 20560, USA.

Animal Protection Institute, Susan Lock, 5894 Southland Park Drive, PO Box 22505, Sacramento, California 95822, USA.

Animal Welfare Institute, Christine Stevens, President, PO Box 3650, Washington DC 20007, USA.

Endangered Species Productions, Phoebe Wray, Director, 175 West Main Street, Ayer, Mass. 01432, USA.

Food and Agriculture Organisation/Advisory Committee on Marine Resources Research Working Party, c/o Dr. Sidney Holt (in charge), Via delle Terme di Caracalle, Roma 00100, Italy.

Fund for Animals, Patricia Forkan, 140 West 57th Street, New York, NY 10019, USA.

General Whale, 9616 McArthur Boulevard, Oakland, California 94605, USA.

Greenpeace, 2007 West 4th Avenue, Vancouver BC, Canada.

Institute for Delphinid Research Ltd, Box N.3531, Nassau, Bahamas.

International Union for the Conservation of Nature and Natural Resources, 1110 Morges, Switzerland.

Project Jonah, Maxine McCloskey, President, c/o Oceanic Society, Building 240, Fort Mason, San Francisco, California 94123, USA.

Project Jonah, Joy Lee, 22 Duff Street, Turramurra 2074, Australia.

Project Jonah, Ross Guy, PO Box 42-071, Orakei, Auckland, New Zealand.

Rare Animal Relief Effort, David Hill, c/o National Audubon Society, 950 Third Avenue, New York, NY 10022, USA.

Save our Whales, Tony Mallin, 6351 North Oakley Ave, Chicago, Illinois 60659, USA.

Save the Dolphins, 1945 20th Avenue, San Francisco, California 94116, USA.

Sierra Club, Michael McCloskey, 1050 Mills Tower, 220 Bush Street, San Francisco, California 94104, USA.

World Wildlife Fund, 1319 18th Street NW, Washington DC, USA.

APPENDIX XIII

GLOSSARY

Aboriginal whaling: whaling carried out for subsistence, the products being consumed by local communities only.

ACMRR: The Advisory Committee on Marine Resources Research of the United Nations' Food and Agricultural Organisation.

Antarctic Treaty: (Act) 1967, enabled the ratification of the measures set out in the Antarctic Treaty signed at Washington in 1959. The Treaty relates to the conservation of the Antarctic flora and fauna.

Antarctic whaling season: The period when the pelagic whalers were permitted to kill whales in the Antarctic. Until 1977, when it was abandoned, it was from December 12th to April 6th.

Ambergris: a substance found in the intestine of the Sperm Whale (found in 3-4 Sperm Whales (Ivashin 1963) out of a hundred killed) or when exuded, floating in the sea. It is used as a fixative in scents.

Baleen: fibrous plates through which baleen whales filter their food.

Biological extinction: complete extinction of the whole species — or so reduced as to preclude the possibility of any recovery.

Bureau of International Whaling Statistics (BIWS): founded in Norway in 1929 and keeps records of each whale caught, its species, sex, length and the region in which it was caught. These records are published annually and can be obtained from: International Hvalfangststatistikk, Postboks 188, 3201 Sandefjord, Norway.

Biomass: a quantative assessment of living things; in this manual it is normally used in the context of total weights of animals or populations.

Blubber: the thick subcutaneous fat of whales.

Blue Whale unit (BWu): a measurement based on the approximate yield of oil which the baleen whales in the Antarctic yielded, ie 1 Blue Whale = 2 Fin or 2½ Humpbacks or 6 Seis. Its usage was abolished in 1972.

Catch per unit effort (CPUE): counting the number of whales caught in relation to the amount of effort it takes to catch them.

Cetacean: derived from the Greek **ketos** meaning a whale. The order comprising whales, dolphins and porpoises.

Commercial extinction: the point at which the population is so reduced as to make its continued hunting uneconomic.

Commercial whaling: whaling carried out for trade purposes.

Density dependent: an influence that is dependent upon a certain density of individuals in order to be fully effective, eg. a limited amount of food (see also p. 66).

Dolphin: loosely used to refer to the small cetaceans with a 'beak'; usually has a dorsal fin.

Ecosystem: the entire inter-dependent flora and fauna of a particular environment.

Flensing: the operation of cutting up a whale.

International Whaling Commission (IWC): secretariat headquarters: Station Road, Histon, Cambridge (see also p. 18).

Krill: a collective term for the small shrimp-like planktonic crustacea. It constitutes the basic diet of the baleen whales in the Antarctic (see also p. *137*).

Maximum sustainable yield (msy): calculated on the assumption that when a population has expanded as much as it can in relation to the availability of space and food, it remains stable, and births will replace the number of individuals which die. But if a proportion is killed, the birth rate may temporarily increase owing to factors such as an increased availability of food for the remainder. The rate which the greatest numbers of whales can be killed without further lowering the population is known as the maximum sustainable yield. Throughout this manual msy = numbers; otherwise, it is specified as msy (weight) (see also p. 68).

Moratorium: "period of legal authorisation to debtors to postpone payment". (Oxford English Dictionary); "suspension of activity, a temporary ban on the use or production of something". (Webster)

Mysticeti: one of the two sub-orders of the order Cetacea. Derived from the Greek: mystax, a moustache. The whalebone or baleen whales.

Odontoceti: one of the two sub-orders of the order Cetacea. The name is derived from the Greek odontos, meaning teeth. Comprises the toothed whales, porpoises and dolphins.

Pelagic: "of, performed on, the open sea — especially sealing and whaling." (Oxford English Dictionary)

Plankton: pelagic flora (phytoplankton) and fauna (zooplankton), floating in the zones near the surface of the sea.

Polygamous: (of Sperm Whales) having more than one mate.

Porpoise: loosely used to refer to the small blunt-headed cetaceans, often without a dorsal fin.

Purse seine: "purse seine fishing uses a long deep wall of webbing (approximately 3 600 feet long and 360 feet deep) to encircle schools of tuna. Speed boats direct the movement of the surface school of porpoises allowing the development of the net at the most advantageous time to capture the tuna. The escape route of the tuna to deep water is cut off by drawing the bottom of the net together (pursing)." (The Whale Book, Endangered Species Productions, USA). Millions of porpoises drown in the tuna nets.

Rorqual: the name rorqual is derived from the old Norwegian word rorhval meaning the grooves that run from just behind the lower lip to the chest. The mouths of the rorquals are not so large as those of the Right Whales, but the grooves, or folds, or pleats allow the floor of the mouth, to drop so increasing the volume of the mouth when feeding. Apart from the grooves, rorquals have the following characteristics which differentiate them from other whales: rorquals are more streamlined than the Right Whales and have flatter heads and short baleen plates. There are six species of rorquals — the Blue, Fin, Sei, Humpback, Bryde's and Minke.

Sonar: navigation by sound.

Spermaceti: whitish fatty substance in the head cavity (case) of Sperm Whales.

APPENDIX XIV

BIBLIOGRAPHY

Becker, J. & Sampson, S., 1974, Net Profit, Project Jonah, USA.

Berzin, A.A., 1971, The Sperm Whale, USA.

Blond, George, 1954, The Great Whale Game, Weidenfeld & Nicholson, London.

Bullen, F., 1898, The Cruise of the Cachalot, Reprint John Murray, 1961, London.

Burton, Robert, 1973, The Life and Death of Whales, Andre Deutsch, London.

Cousteau, J-Y., 1972, The Whale, Cassell, London.

Discovery Reports, 1938, London. Most of the British research into Antarctic whales is published in these reports.

Fichtelius, Karl-Erik & Solander, Sverre, 1973, Man's Place: Intelligence in Whales, Dolphins and Humans, Gollancz, London.

Friends of the Earth, 1972, Whale Campaign Manual No. 1, FoE Ltd, London.

Friends of the Earth, 1974, The Position of Whaling in the Japanese Economy, Project Jonah/Friends of the Earth, London.

Gambell, Ray, 22/6/72, Why All the Fuss about Whales?, New Scientist, London.

Haley, Nelson Cole, 1950, Whale Hunt, Travel Book Club, London.

Harmer, S.F., 1928, History of Whaling, Proc. Linn. Soc., London, **140**:51-95.

Harmer, S.F., 1931, Southern Whaling, Proc. Linn. Soc., London, **142**:85-163.

Harrison Matthews, L., (ed), 1968, The Whale, George Allen & Unwin, London.

Hinton, M.A.C., 1925, Report on the Papers left by the late Major Barrett-Hamilton, relating to the whales of South Georgia, Crown Agents for the Colonies, London: 57-209.

Housby, Trevor, 1971, The Hand of God, Abelard-Schuman, London.

International Whaling Commission Reports

International Whaling Statistics, 1932, Oslo.

Koers, A.W., 1973, International Regulation of Marine Fisheries, Fishing News (Books) Ltd., London.

Lack, D., 1954, The Natural Regulation of Animal Numbers, Oxford.

Laws, R.M. & Purves, P.E., 1956, The Ear Plug of the Mysticeti as an indication of the Age with special reference to the North Atlantic Fin Whale, Nork, Havlfangsttid:45 (8):413-425.

Lilly, J.C., 1962, Man and Dolphin, Gollancz, London.

Lockyer, C., 1972, The Age of Sexual Maturity of the Southern Fin Whale, Journ. Cons. Int. Explor. Mer. vol 34 No. 2:276-294.

Lyubimova, T.G., Naumov, A.G., Lagunov, L.L., 1973, Prospects of the Utilisation of Krill and Other Non-conventional Resources of the World Ocean, J. Fish. Res. Bd. Canada 30 (12 pt 2).

Mowat, Farley, A Whale for the Killing, Heinemann.

National Marine Fisheries Service/National Oceanographic and Atmospheric Administration, 1974, Report to Secretary of Commerce, Administration of the Marine Mammal Protection Act of 1972.

Mackintosh, N.A., 1965, The Stocks of Whales, Fishing News (Books) Ltd., London.

Melville, Herman, 1881, Moby Dick, London, Numerous reprints.

Mitchell, Edward D., 1973, The Status of the World's Whales, Nature Canada Vol No. 2 No. 4:9-25.

Nayman, Jaqueline, 1973, Whales, Dolphins and Man, Hamlyn, London.

Morris, K.S., 1966, Whales Dolphins & Porpoises, Univ. Calif. Press., Los Angeles.

Norsk Hvalfangst Tidende 1912-1968, Oslo. (The Norwegian Whaling Gazette).

Ommanney, F.D., 1971, Lost Leviathan, Hutchinson, London.

Purves, P.E. & Mountford, M.O., 1959, Ear Plug Laminations in relation to the Age Composition of a Population of Fin Whales, Bull. Brit. Mus. Nat. Hist. 5 (6):125-154.

Rice, D.W. & Wolman, A.A., 1971, Life History and Ecology of the Gray Whale, USA.

Robertson, R.B., 1956, Of Whales and Men, Macmillan, London.

Ruspoli, Mario, 1972, Les Hommes de la Baleine, Offidoc, Paris.

Scheffer, Victor B., 1969, The Year of the Whale, Charles Scribner's Sons, New York, and Penguin, Harmondsworth, Middlesex.

Scientific Report of the Whales Research Institute, Tokyo, 1948.

Scoresby, W., 1820, An Account of the Arctic Regions with a History and Description of the Northern Whale Fishery, Edinburgh, (Reprint David & Charles, Newton Abbot).

Slijper, E.J., 1962, Whales, London.

Small, George L., 1971, The Blue Whale, Columbia Univ. Press., New York and London.

Venables, B., 1969, Baleia! Baleia!, Bodley Head, London.

Wise, Terence, 1970, To catch a Whale, Bes. London.

Wood, Forest G., 1973, Marine Mammals and Man, Robert B. Luce Inc., New York.

WHALE MANUALS & TEACHING KITS

The Whale Book: a Conservation Manual, Teaching Tool Source Book, 1974, Endangered Species Productions, PO Box 472, Prudential Centre Station, Boston, Mass. 02116, USA.

Whale Campaign Manual, Friends of the Earth and Project Jonah, Australia, 1973.

Project Jonah, Teaching Kit on Whales.

Friends of the Earth Ltd. (FOE) is an environmental pressure group funded by voluntary contributions. It has over 170 affiliated groups in Britain and is part of a world-wide federation of similar organisations. FOE actively pursues campaigns on the control of packaging, the protection of endangered species, food, energy, materials use, recycling, transport and land-use policy. It is associated with the environmental research charity, Earth Resources Research Limited.

If you would like to join Friends of the Earth to campaign for better use of the Earth's resources, or would like more information, please complete the form below:

- -

To: FRIENDS OF THE EARTH LTD,
    9 POLAND STREET, LONDON, W1V 3DG.
    (Telephone 01-434 1684)

I would like more details of Friends of the Earth   ☐
(please enclose s.a.e.)

I would like to register as a Friend of the Earth   ☐

Name_____

Address_____

_____

I have enclosed a registration fee of £3
(registration is valid for 12 months)